Alpheus S. Packard

The Anatomy, Histology and Embryology of Limulus

Polyphemus

Alpheus S. Packard

The Anatomy, Histology and Embryology of Limulus Polyphemus

ISBN/EAN: 9783337403737

Printed in Europe, USA, Canada, Australia, Japan

Cover: Foto ©berggeist007 / pixelio.de

More available books at **www.hansebooks.com**

THE ANATOMY, HISTOLOGY, AND EMBRYOLOGY

OF

LIMULUS POLYPHEMUS.

By A. S. PACKARD, Jr., M.D.

BOSTON:
PUBLISHED BY THE SOCIETY.
1880.

The Anatomy, Histology and Embryology of Limulus Polyphemus.

By A. S. Packard, Jr., M.D.

Since the publication of my first paper on the development of the horse-shoe or king crab (*Limulus polyphemus*), in the Memoirs of this Society,[1] I have, as opportunity allowed, made additional observations on the development of the larva, and also on the histology of the different organs, and especially the brain. In making the microscopic sections of the embryos and for a series of sections of a brain, the latter of which were unstained, I am indebted to Professor T. D. Biscoe. For mounting some of these sections for study, I am indebted to Dr. C. B. Johnson of Providence, R. I., who also kindly cut, stained, and mounted preparations of the digestive canal. Within the past year I again returned to a study of the brain, using the methods of staining employed by German observers, Dietl and Krieger, also by Mr. E. T. Newton. The sections of the brain were cut and stained, as also those of the eyes, parts of the stomach and rectum, kidneys and liver, etc., by Mr. Norman N. Mason, of Providence, R. I., who kindly devoted a great deal of time to the work. To his unusual skill and delicacy of manipulation, I am indebted for a large number of preparations much better than I could have made myself, and which have been of most essential aid in preparing this paper; so that portions of the histological part of this paper, especially that on the structure of the eyes, are really joint productions with Mr. Mason, as we together examined the preparations.

Position of Limulus among Arthropoda.

The researches of M. Alphonse Milne-Edwards on the anatomy of Limulus, proved that this animal, so far from being a genuine normal crustacean, is either the type of a group equal to all the other Crustacea, namely a sub-class of Branchiata; or, as several authors contend, should be regarded as the representative of a distinct class of Arthropoda.

Before arguing what we now believe to be the true position of Limulus and the allied fossil forms, including the Trilobites, let us take a review of the different opinions of the leading zoologists who have done special work on the animal. The titles of their work will be found in the bibliographical list at the end of this paper.

[1] Memoirs Bost. Soc. Nat. Hist., II, 155-202.

Straus-Dürckheim was the first author to remove the genus Limulus from the Crustacea, and to regard it as the type of a distinct order of Arachnida, which he called Gnathopodes. In his memoir, published in 1829, according to Van der Hoeven's statement, Straus characterized the Arachnida by the disposition of the feet arranged in a circle around an interior cartilaginous sternum, and by the absence of antennae. Van der Hoeven, in 1838, remarks that the branchiae are the principal characters of Crustacea, as insisted upon by Latreille and Milne-Edwards, who placed Limulus in this class; therefore Limulus should belong with these animals, and he shows that there are other characters which separate Limulus from the Arachnida, and which ally them with the Crustacea. These are the compound eyes, the position of the stomach in the front of the cephalothorax, "while it is contained in the abdomen of Arachnida." He then says: "But whether we place the Limuli among the Crustacea, or with the Arachnida, they should always form a distinct order for themselves alone, which, in the actual state of our knowledge, is far from all the other orders of these two classes." Afterwards, in 1846, in his Handbook of Zoology, and again in the second, English edition of 1856, he placed the Poecilopoda as the first order of Crustacea, referring, however, to their resemblance to Arachnida.

In 1871, Dr. A. Dohrn, in his Untersuchungen über den Bau und Entwicklung der Arthropoden, concluded that Limulus, Eurypterida and Trilobita should be united under a common name, Gigantostraka, as originally proposed by Haeckel, in his Morphologie, for the Eurypterida alone; and that they should be placed near the Crustacea.

Most if not all the other leading zoologists, while recognizing the aberrant characters of the Limuli, have left them among the Crustacea, though in 1834 H. Milne-Edwards established a subclass (Xiphosura) for the group; this group being equivalent to any one of several other subclasses of Crustacea which he enumerates. For the views we held previous to the publication of H. Milne-Edwards' memoir, we would refer the reader to our Memoir on the development of Limulus, published in March, 1872.

In October, 1871, the following views of M. Édouard Van Beneden[1] were published : "L'étude du développement embryonnaire de ces animaux et de leurs caractères anatomiques m'a conduit aux conclusions suivantes que je puis formuler dès à présent:

I. Les Limules ne sont pas des Crustacés; ils n'ont rien de commun avec les Phyllopodes, et leur développement embryonnaire présente les plus grandes analogies avec celui des Scorpions et des autres Arachnides, dont on ne peut les séparer. Dans le cours de leur développement embryonnaire, on ne distingue aucune des phases caractéristiques du développement des Crustacés, et il ne peut être question de distinguer dans le cours de ce développement embryonnaire, ni phase nauplienne, ni phase cyclopéenne.

II. L'analogie entre les Limules et les Trilobites, et l'affinité qui relie entre eux ces deux groupes, ne peut être un instant douteuse pour celui qui a étudié le développement embryonnaire de ces animaux. Les lois de développement sont les mêmes chez les Trilobites et les Xiphosures, et l'analogie entre les jeunes Trilobites et

[1] Journal de Zoologie. Par Paul Gervais. Tom. I, p. 42, 1872. Paris.

les jeunes Limules est d'autant plus grande, qu'on les considère à une époque moins avancée de leur développement. A l'examen de jeunes Limules, MM. Packard et Woodward ont été frappés de ces analogies.

III. Les Trilobites, aussi bien que les Euryptérides que les Poecilopodes, doivent être séparés de la classe des Crustacés et former avec les Scorpionides et les autres Arachnides un rameau à part, dont l'origine est encore à déterminer."

In November, 1872, A. Milne-Edwards, in his beautiful memoir on the anatomy of Limulus, claimed that the central nervous system resembled that of Arachnida, and was surrounded by arterial coats, and that the brain supplied no limbs with nerves. His conclusions are stated in the following extracts : " L'après les faits que je viens de passer en revue, on voit que le système nerveux de la Limule diffère beaucoup de celui de tout autre animal articulé, et resemble moins à celui des Arachnides qu'à celui des Crustacés. Chez les premiers, les ganglions cephalothoraciques sont tellement serrés entre eux que le pertuis ménagé au milieu du collier œsophagien, pour le passage du tube alimentaire est d'une petitesse extrême, et qu'en arrière de cette masse médullaire, les deux moitiés de la chaine nerveuse sont réunies entre elles dans toute leur longeur, au lieu d'être attachées l'une à l'autre par des commissures ganglionnaires seulement. Chez les Crustacés, on rencontre souvent une disposition analogue à celle des Limules. Mais la coalescence des ganglions cérébroïdes et des ganglions postbuccaux n'est jamais portée aussi loin, et c'est en général entre ces deux systèmes des centres nerveux que les connectifs sont le plus allongés. Chez les Limules, au contraire, ces connectifs sont remarquablement courts, tandis que ceux situés à la partie antérieure de la région abdominale sont fort longs. Il est aussi à noter que le système ganglionnaire viscéral, dont M. Blanchard a tiré des caractères anatomiques pour la distinction des Insectes, comparés aux Myriopodes et aux Arachnides, présente chez les Limules une disposition qui n'a encore été observé nulle part ailleurs. Ces particularités anatomiques viennent donc à l'appui de l'opinion que j'ai déjà émise, relativement à la nécessité de séparer ces animaux des autres Articulés, et d'en former une classe particulière, sous le nom de Merostomata, classe très-voisine, d'ailleurs, des Arachnides." He then states, in considering the external anatomy, that it is not only by their internal organization that the Limuli differ from the Crustacea and approach the Arachnides, without, however, being confounded with them ; for there are also in the general conformation of the Merostomata and the Scorpions, resemblances which seem to indicate in all these Entomozoa a community of primordial type.

The external characters which separate the Limuli from all other articulated animals are the absence of any preoral appendages, Milne-Edwards having shown that the nerves to the first pair of feet do not arise, as Van der Hoeven and Owen claim, from the brain, but from the oesophageal collar. To use Edwards' own words : " J'en conclus que, chez les Limules, il y a absence complète d'appendices frontaux, et ce caractère les distingue des Arachnides aussi bien que de tous les autres animaux articulés de la période actuelle."

Finally, he remarks that if the Limuli are not Crustacea, neither are they Arachnida. " They are distinguished, the latter not only by their mode of respiration, but by the existence of compound eyes, the absence of frontal appendages, the continuous prolongation of the ventral appendages on the adjacent part of the abdomen, and by several other organic characters. They are distinguished from all other articulated animals by the disposition of their circulatory system, and consequently, in spite of the small number of species of this group, the zoologist should consider them as constituting a particular class intermediate between the Crustacea and Arachnida. He claims with Mr. H. Woodward, that the fossil Pterygoti and Eurypteri should be united with the Limuli, under the name of Merostomata. · Milne-Edwards then adds that "the Merostomata were contemporaries of the Trilobites, and there seems to be between these two groups, not only very strong resemblances, but intermediate forms which establish the passage from one to the other. Some authors have thought it useful to unite them under a common name. This seems to me at least too premature, because we know nothing of significance on the subject of the appendicular system of Trilobites, and we cannot pronounce legitimately on this question ; but it should be taken into consideration, that it seems very probable that the Trilobites differ from the Crustacea properly so-called, as we have seen the Merostomata differ from them, and that they should likewise constitute a particular class in the great natural division of Entomozoa."

In November, 1873,[1] in the light of A. Milne-Edwards' researches, I stated that "I should no longer feel warranted in associating Limulus and the Merostomata generally with the Branchiopoda, but regard them as perhaps forming with the Trilobites a distinct sub-class of Crustacea. In a second notice in the same Journal for December, 1879, I proposed the name Palaeocarida, for the sub-class; these comprising the Merostomata and Trilobites. We also proposed the term Neocarida for the remaining sub-class of normal Crustacea.

In 1874 Gegenbaur, in his Grundriss der Vergl. Anatomie, divides the living Branchiata as opposed to the Arthropoda Tracheata, in two divisions: I. Crustacea, II. Poecilopoda.[2]

As regards the relations of the Merostomata to the Arachnida let us examine them and inquire whether they are not rather those of analogy, than of affinity. It is not necessary, in view of what has been published, for us to restate the essential anatomical characteristics of Limulus. The relations of the viscera to the body wall, and of the appendages may be seen by our figures in Plates I and II.

The resemblances to the Arachnida in general, and the scorpion in particular, have been supposed to consist (1) in the want of antennae, and (2) the form of the central nervous system, as well as (3) the mode of development, while (4) the branchiae of Limulus have been homologized with the pulmonary sacs of spiders.

It should be borne in mind, however, that the Arachnida are a sub-class of Tracheata, with no antennae to be sure, but with two pairs of post-oral appendages, the mandibles and maxillae, which are constructed on the hexapodous type, and are also built upon the same plan of structure as the mouth-appendages of Myriopoda ; so close indeed are the homologies between the Hexapoda, or insects proper, and the Arachnida and Myriopoda,

[1] Farther observations on the embryology of Limulus, with notes on its affinities. Amer. Naturalist, November, 1873.

[2] If we substitute for the term Poecilopoda, which applies only to the sub-order of which Limulus is the type, the term Palaeocarida, and regard Gegenbaur's Crustacea as equivalent to my Neocarida; this would express my views as to the relations of the two sub-classes. This makes the terms Crustacea and Branchiata synonyms from my point of view.

all breathing by trachese, excepting the few species which have no breathing organs at all, that it seems most advisable to retain them as sub-divisions or sub-classes of the class of insects or Tracheata.

There is little in common between the mouth-parts of Limulus and those of the Arachnida, either in their form or grouping; moreover, the mouth-parts of Limulus are not differentiated from the other cephalothoracic appendages. The six pairs are alike; morphologically true gnathopods; and in the embryo arise simultaneously; in the Arachnida, the two pairs of mouth-parts are, in adult life, quite different from the eight legs, and are soon differentiated in early embryonic life. Limulus resembles the Arachnida in the want of antennae, but so important are the differences in the mouth-organs and legs, that it seems a violation of the principles of classification to associate together the two types within the limits of the same class.

The second Arachidan feature claimed by authors to exist in Limulus is the alleged similiarity in the form of the nervous system to that of the Arachnida, especially the scorpions and spiders. The oesophageal collar of the horse-shoe crab has been homologized with the thoracic ganglionic mass of Arachnida, and the brain of Limulus has been likened to that of the spiders and of the scorpions.

The brain of Arachnida has heretofore been supposed to be a single pair of ganglia, and to send nerves not only to the simple eyes, but also to the first pair of mouth appendages. If this view is correct, as all who have studied the adult Arachnids agree, then the brain of Limulus is not homologous with the arachnid brain (supra-oesophageal ganglion), as it supplies only the eyes, sending no nerves to the anterior gnathopods. As will be seen farther on (Plate 4, fig. 7 gn), the first pair of gnathopods is supplied in the larva directly from an independent pair of ganglia. Very recently, however, Mr. Balfour [1] has proved that the so-called supra-oesophageal ganglion or brain of the spider is formed of two pairs of ganglia which at first are quite distinct, as shown by his section of the embryo spider. Mr. Balfour concludes that "the evidence which I have got that the cheliceres are true postoral appendages, supplied in the embryo from a distinct postoral ganglion, confirms the conclusions of most previous investigators, and shows that these appendages are equivalent to the mandibles, or possibly the first pair of maxillae of other Tracheata."

In either case then, whether the brain of Arachnida is a single pair of ganglia, supplying the cheliceres (or mandibles), as well as the ocelli or two pairs of consolidated ganglia, the brain of these Arthropods can scarcely be homologous with the brain of Limulus.

Moreover, the position of the brain in relation to the thoracic ganglionic mass of Arachnida is quite different from that of Limulus; in the former animals, judging from Blanchard's beautiful and accurate plates, and our own examination of the brain of the scorpion, it is invariably situated in a plane parallel to and much above the thoracic mass, and separated by long slender commissures; while the brain of Limulus is situated on the same plane as the oesophageal collar, in fact, closing up the front of what would otherwise be an open ring or collar.

[1] Notes on the development of the Araneina. By F. M. Balfour. Quarterly Journal of Microscopical Science, April, 1880, pp. 176, 185, 189, Pl. xxi., fig. 21.

The thoracic ganglionic mass of the Arachnida is likewise not homologous with the central cephalothoracic nervous system of Limulus. The thoracic mass in the former type sends off nerves to the maxillae, or second pair of mouth-appendages, and also to the four pair of limbs, and from this mass the abdomen, including the spinnerets (in our view morphologically limbs), is supplied with nerves; there being no ganglia in the abdomen of any spiders (Araneina) as yet known. On the contrary, the oesophageal collar of Limulus supplies the nerves for the six cephalothoracic appendages alone (and this seems strong proof that these gnathopods should be regarded as either mouth-parts alone, or partly mouth appendages, and partly thoracic appendages), while there is a chain of six ganglia in the abdomen. Here, however, it should be borne in mind that in the scorpions there is a chain of abdominal ganglia, so that in this respect there is an interesting analogy between Limulus and the Pedipalpi. So far, however, as concerns the brain and thoracic mass, there seems to be a lack of homology in the two types of nervous system of Limulus and Scorpio.

In the mode of early development, Limulus resembles the Arachnida, but also in the embryonal membranes the insects, while it also recalls the development of certain Crustacea, notably Apus, as we attempted to show in our first memoir.

The fourth point of comparison, i. e., between the gills of Limulus and the pulmonary branchiae of spiders seems far-fetched. The gills and mode of respiration of Limulus are thoroughly crustacean, the gills being certainly not homologues of the "lungs" of the air-breathing spiders, which are tracheal sacs, formed by modified tracheae, and opening externally by stigmata.

From any point of view, developmental, anatomical or physiological, the relations of Limulus and its fossil allies to the Arachnida seem purely those of analogy, the fundamental differences being such as characterize and separate the Tracheate from the Branchiate Arthropods; the differences are so fundamental as to suggest the idea that the two types probably had a different origin, i. e. from some vermian ancestors.

In order to epitomize the differences and resemblances between the Merostomata and Arachnida, we have prepared the following tabular view:

COMPARISON OF THE MEROSTOMATA WITH THE ARACHNIDA.

Arachnida.	Merostomata (Limulus).
Head in adult soldered to thorax.	Head separate from hind body.
No compound eyes.	Compound eyes.
No antennae or morphological equivalents.	No antennae or morphological equivalents.
Mandibles on hexapodous type.	Only their morphological equivalents (gnathopods).
Maxillae with a palpus, on hexapodous type.	" " "
Four pairs of thoracic legs on hexapodous type.	No true thoracic legs; the gnathopods representing the mouth-parts and possibly the thoracic legs.
No functional abdominal legs, the spinnerets being, however, modified legs.	Six pairs of swimming respiratory legs, on the Crustacean type.
Digestive canal on hexapodous type with a voluminous liver, and urinary tubes.	Digestive canal on Crustacean type, with a voluminous liver, but no urinary tubes.
Brain formed of two pairs of ganglia supplying eyes and mandibles.	Brain formed of a single pair of ganglia, supplying eyes alone, and free from the suboesophageal ganglion in embryo and adult.
Maxillae and thoracic legs supplied from a concentrated postoesophageal ganglionic mass.	Gnathopods supplied from a concentrated ganglionic oesophageal ring.
No abdominal ganglia in spiders, but present in scorpions.	Six abdominal ganglia, much as in Crustacea.

Turning now to the relations of the Merostomata to the normal Crustacea, we may inquire whether the former belong to the class of Crustacea, or should form the type of a distinct class. The latter view is that proposed by A. Milne-Edwards; and a number of zoologists have adopted this view.

The facts that seem to us to point to the crustacean nature of Limulus and its allies are: (1) the nature of the branchiae, those of Limulus being developed in numerous plates overlapping each other on the second abdominal limbs; those of the Eurypterida being, according to H. Woodward, attached side by side, like the teeth of a rake; while the mode of respiration, as seen on plate 1, is truly crustacean; (2) the resemblance of the cephalothorax of Limulus to that of Apus; (3) the general resemblance of the gnathopods to the feet of the Nauplius or larva of the Cirripedia and Copepoda; (4) the digestive tract is homologous throughout with that of Crustacea, particularly the Decapoda, there being no urinary tubes as in Tracheata; (5) the heart is on the crustacean type as much as on the tracheate type, and the internal reproductive organs (ovaries and testes) open externally, at the base of and in the limbs, much as in Crustacea.

The resemblances and differences between the normal Crustacea (Neocarida) and the Palaeocarida (Merostomata and Trilobita) are shown in the following tabular view :—

COMPARISON OF NORMAL CRUSTACEA (NEOCARIDA) WITH LIMULUS AND OTHER PALAEOCARIDA.

Neocarida.	Palaeocarida.
Integument solid and calcareous, or thin and chitinous.	Integument usually chitinous.
Usually in higher forms a cephalothorax, but in Phyllopods no genuine cephalothorax distinct from the abdomen.	Head and abdomen alone; no thorax except in trilobites.
Eyes of normal form, rods and cones present, but no corneal lenses.	Eyes with no rods and cones, but corneal lenses.
Two pairs of antennae.	No antennae, either functional or morphological.
Mandibles normal.	No functional mandibles = gnathopods.
Maxillae normal.	No functional maxillae = gnathopods.
Maxillipeds normal.	No functional maxillipeds = gnathopods.
Gills on thoracic feet, or thoracic or abdominal feet themselves broad and thin, and serving as gills.	Gills on the abdominal feet.
Abdominal feet biramous.	Abdominal feet biramous.
Heart polygonal or tubular.	Heart tubular, as in many Neocarida except Decapoda.
Digestive canal with its three subdivisions of fore-, mid- and hind-gut.	Digestive canal homologous with that of most higher Crustacea.
Nervous system with a brain sending nerves to the antennae and eyes.	Nervous system with brain supplying eyes alone—first pair of gnathopods supplied from oesophageal collar, in larva from suboesophageal ganglion.
Oviduct opening at base of middle thoracic feet ; male outlet at base of 5th thoracic feet.	Oviduct and male outlet situated at base of first abdominal feet.
Metamorphosis often complete.	Metamorphosis absent, or partial.
Nauplius in some forms.	No Nauplius.
Zoea in Decapods.	No Zoea.

The difficulties which stand in the way of associating the Merostomata (throwing out the Trilobites for the sake of clearness of statement) with the Crustacea, are : (1) the nature of the limbs, and the absence of the pairs of antennae; but it may be observed that in the undifferentiated gnathopods of Limulus we have a parallel in the larval

Cirripedia and Copepoda, where what ultimately become antennae and mandibles are swimming feet; and in the zoea of Decapods, in which two pairs of antennae exist, and the temporary swimming feet ultimately become maxillae and maxillipedes; (2) the unique relations of the inferior blood system to the central nervous system (the brain and certain nerves alone excepted); and (3) the peculiar nature of the eyes of the Merostomata and Trilobites, which are constituted on a type peculiar to themselves.

Under all these circumstances, it may be claimed, as has been done by A. Milne-Edwards, that the Merostomata should form a distinct class of Arthropoda. It should be borne in mind, however, that M. A. Milne-Edwards believes that a second class of Arthropods should be formed to receive the Trilobites. Taking all the facts into consideration, we should propose that the Merostomata and Trilobites should together form a subclass of Crustacea (i. e., Branchiate Arthropods) standing parallel to, and as the equivalents of, all the other Crustacea, the two groups being parallel and equally important branches of the same genealogical tree.

It should be borne in mind that the Palaeocarida are a generalized or synthetic type; Limulus is, so to speak, a subzoea, the cephalothorax having been differentiated from the abdomen and prematurely developed, with the gills of a normal crustacean; having the primitive appendages of a nauplius, and the compound eyes superficially like those of a zoea, but on an elementary, prematurely developed type; while the circulatory system is of a high order, and the nervous system well developed, though the brain is constituted on a simple plan, quite unlike that of the higher Crustacea, and probably the Crustacea in general. The subclass of Palaeocarida apparently bears very much the same relation to the subclass Neocarida, as the subclass Elasmobranchii or Ganoidea do to the Teleostean fishes; as in these early synthetic forms certain organs are prematurely developed, while the skeleton and other parts are in a more or less embryonic or larval condition. They abounded most in the Palaeozoic ages, dying out in part, with but a few survivors: such was the case with the Palaeocarida. Under these circumstances we see no more reason for removing the Merostomata and Trilobita from the class of Crustacea, than to consider the Elasmobranchii or Ganoids as independent classes of Vertebrates, or the Arachnids or Myriopoda, as independent classes of Arthopoda.

Regarding, then, the Palaeocarida as an early offshoot of the Crustacean or Branchiate Arthropod tree or stem, we would venture to present the classification on the following page, as proposed in 1879, in our little school book, "Zoology."

The Neocarida may be characterized briefly as genuine Crustacea with two pairs of antennae, biting mouth-parts and ambulatory or swimming thoracic feet; mostly modern types. The Palaeocarida, on the other hand, have the cephalothoracic appendages in the form of foot-jaws, rather than true jaws; no antennae, the brain supplying the compound eyes and ocelli alone; the nerves to the cephalothoracic appendages sent off from an oesophageal ring or collar; and the nervous system, with the exception of the brain, ensheathed in a ventral system of arteries; they are mostly palaeozoic types.

The close homologies between the Merostomata and Trilobita were discussed in our first memoir. At that time (p. 184), we advocated the view that the cephalothoracic limbs of the Trilobites must have been jointed, rounded rather than foliaceous, and ambulatory in function, and inclined to the views of Mr. Billings as to the nature of what he regarded the appendages of the Asaphus described by him in 1864. Since then the researches of Mr. C. D. Walcott[1] on sections of Trilobites seems to have satisfactorily proved that Trilobites have rounded, jointed ambulatory appendages developed from the head and possibly from the thorax. His observations, though from the nature of the case in some respects imperfect, have set at rest the question as to whether these extinct Palaeocarida had rounded, jointed limbs, though much yet remains unproved as to the homologies of these limbs with those of the Merostomata. It also appears that the hard parts of the eyes of Trilobites are directly homologous with those of Limulus, as we attempt to show hereafter in this paper.

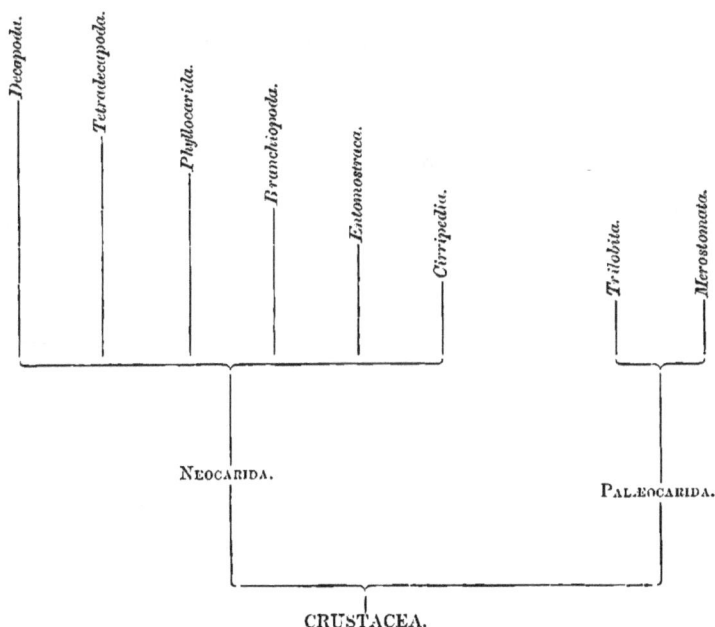

As to the general homologies of the body of Limulus, it seems to us that the facts presented further on confirm the position we have always taken, i. e., that there are no true antennae in Limulus; that the gnathopods are mostly modified mouth-parts, the last pair possibly representing a pair of thoracic feet; that the fore region of the body corresponds to the cephalothorax of the Decapoda or of a

[1] Preliminary notice of the Discovery of Natatory and Branchial appendages of Trilobites, and additional evidence upon the same. Twenty-eighth Annual Report, New York State Museum of Natural History, December, 1876. Notes on some sections of Trilobites from the Trenton Limestone, Sept. 20, 1877. See also Ann. Rep. N. Y. Mus. Nat. Hist., March, 1879.

Nebalia, and that the posterior region is truly an abdomen, the spine of Limulus
being simply the last body-segment, or ninth abdominal arthromere, as the history
of the embryonic development of this segment proves. It then follows that the
abdominal respiratory feet are, for example, homologues of the broad respiratory
abdominal appendages of Isopoda. The view of Mr. Woodward, that what we regard
as abdomen represents in part the thorax, or the opinion of Owen and Huxley that the
spine represents the abdomen, and that what we call the abdomen is the thorax, in
part at least, is, it seems to us, not based on sound induction.

HISTOLOGY OF THE INTERNAL ORGANS OF THE ADULT LIMULUS.

Histology of the digestive system. The general form of the digestive canal is
seen in plate 3, fig. 1. The large mouth-opening is situated between the third to
fifth pairs of limbs. The oesophagus is very long, and directed very obliquely forward
and upwards from the mouth, entering the large crop or proventriculus at an angle
to the general course of the latter, which is full and large, projecting anteriorly
over the end of the oesophagus. It curves over backwards, growing smaller posteriorly,
projecting above slightly over the beginning of the stomach or mid-gut. What we call
the crop, is the "cardia" of Van der Hoeven, and the "cardiac end of the stomach"
of Owen and A. Milne-Edwards. Communication with the chyle-stomach is effected
by the large internal projection in the form of a truncated cone (plate 3, fig. 1, *cone*),
by which the food, when partially digested, is strained, and passes from the proventriculus
into the true stomach. The latter, externally, seems to form the beginning of the
intestine, and extends from the base of the proventricular projection backwards as
far as the first pair of biliary ducts; its histology is quite different from that of the
proventriculus and its posterior conical process.

The beginning of the intestine is indicated externally by a slight contraction just
before the origin of the anterior of the two pairs of biliary ducts. These are placed
far apart by a distance nearly equal to twice the thickness of the intestine. The
hind gut is divided into the intestine and rectum. The intestine is straight, and
of uniform thickness as far as the beginning of the rectum, which is swollen, owing to the
large rectal folds within.

On laying open the digestive canal of specimens collected in the winter, it is found
to be filled with a jelly-like substance, which on examination proves to be the lining
of the canal, which has been molted, and has undergone partial digestion.

Examining the inner walls of the digestive canal, and studying its histology, we
find that there are three fundamental layers composing the canal, extending from the
mouth to the vent. There are, beginning on the outside, (1) the muscular layer,
(2) the mucous or epithelial layer, and (3) the chitinous layer. The muscular
layer is made up of longitudinal muscles, the fibres striated, with scattered small bundles
of transverse striated fibres, some of these isolated from the outer layer of longitudinal
muscles and passing through the epithelial tissue.

The second or epithelial layer is thick, composed of pavement epithelium, arranged in
fibrous masses or bundles, somewhat like muscular tissue. The nuclei are large and

conspicuous where the preparations have been stained with haematoxylin;[1] the cell walls are difficult to distinguish with a one-fifth objective. The pavement epithelium fills the spaces between the folds of the oesophagus and crop (or fore gut), and is succeeded by a single layer of columnar epithelium, which looks like a delicate ruffle, edging the folds, and lying between the pavement epithelium and the chitinous lining of the canal. The chitinous layer is very finely laminated, the laminae being parallel for the most part to the indentations and projections of the folds and the teeth of the fore-gut, showing plainly that it is secreted by the layer of columnar epithelium. Cross sections of the larva, after hatching, through the fore-, mid-, and hind-gut, when the appendages and internal organs have assumed their definite shape, show that the intestine then consists of only two layers, the muscular, which is comparatively thin, and the layer of columnar epithelium (plate 5, figs. 7, 7a), which rests directly upon the muscular layer, and consists of long cells projecting irregularly into the cavity of the canal. It would thus appear that the thick layer of pavement epithelium and of chitine is not developed throughout the intestine, until some time after hatching. Indeed, it is known that the larva lives for a long time, even months, after hatching, before it takes much, if any, food.

Returning to the oesophagus; it is seen to be lined with a pale yellowish chitinous layer gathered into about eight large deep folds. Plate 5, fig. 5, illustrates the structure of two of these folds and part of the adjoining ones. The muscular fibres are not represented. The cells 4, 4a, of the pavement epithelium (pe) are round or oval, with a large, distinct, dark nucleus; their walls are difficult to define. The projecting lobes consist of columnar epithelium, with large nuclei, much more distinct than in the pavement epithelium; the basal half of the cells are dark, being filled compactly with granular matter enclosing the nuclei, while on the outer half the cells are transparent; plate 5, fig. 3, 3a, represents these cells enlarged. The lobes are hollow, leaving a clear space, as shown in figure 5; the lobes are unequal in form and size, those figured being situated near the posterior end of the oesophagus. The columnar epithelium is succeeded by the chitinous layer (ch), which is finely laminated, the laminae corresponding to the direction of the lobes.

The crop or proventriculus consists of three parts; in the most anterior division the chitinous folds, continuous with those of the oesophagus, are large and irregular and extend vertically upwards, until they bend backwards suddenly at right angles to form the rows of thick, solid teeth lining the second or middle and larger part of the crop. These teeth are arranged in five sets of rows, each set or series consisting of three rows, and two series of two rows, the two latter sets situated on the under or ventral side of the stomach, and arranged on each side of the three-rowed series. The teeth in each row are nearly uniform in size, are transverse, being flattened antero-posteriorly. In the three-rowed series, especially on the ventral side, the teeth of the middle of the three rows are larger than those of the row on each side. There are about 225 well marked teeth in this division of the crop, those at either end of the rows being small and sometimes double.

---

[1] I am much indebted to Dr. C. B. Johnson, of Providence, for kindly cutting, staining, and mounting some excellent preparations of the oesophagus, crop, and intestines, stained both with haematoxylin and carmine.

The minute structure of the vertical folds of the first or anterior division of the crop may be seen at plate 5, figs. 1. 2, where the relations between the muscular, epithelial and chitinous layers are shown. The limits between the longitudinal striated muscular layer (m) and the epithelial layer are clear and well marked, the bundles of pavement epithelium (pe) running at right angles to and abutting directly on the muscular layer. The pavement epithelium is also, in slices stained with haematoxylin, clearly demarked from the columnar epithelium (ce) by its pale lilac tint, the latter staining brownish and contrasting well with the purple-stained chitine, which is finely laminated, the lines of deposition being waved, the points of the waves under a low power appearing like fine lines passing inward near but not quite to the free edge of the tooth, the margin of the chitinous layer remaining unstained and pale yellowish. Fig. 2 represents the small central tooth 1x, still more enlarged, showing the lines of growth of the chitin, and the ruffle of columnar epithelium, indicated by the row of large nuclei bordering the margin of the lobes of the columnar epithelium layer.

In the teeth of the middle region, which as we said, number some 225, the columnar epithelium is wanting, though the corresponding tract is yet stained pale brown by haematoxylin or deep crimson by carmine, but the cells are of the same nature as in the adjoining pavement epithelium; it is also not scalloped, but the layer of chitine is much thicker than elsewhere in the digestive canal.

The proventricular cone or tube has internally about thirteen unequal chitinous folds, continuous with and like those of the oesophagus, five large folds alternating with eight smaller ones. The folds are yellowish, and project ruffle-like at the end, contrasting in structure and color with the whitish exterior of the cone or strainer. An examination of the cellular structure of the interior lining of this tube, shows that it has a chitinous lining continuous with that of the crop, and which stops at the ruffle-like extremity of the tube; this chitinous layer is succeeded within by a ruffle-like layer of columnar epithelium. like that in the fore part of the crop. The chitin is entirely wanting in the papillose exterior of the tube, while the layer of columnar epithelium is deep, the cells being very long and slender. The structure, then, of this tube is externally like that of the stomach walls as described below, while internally it is histologically an extension of the structure of the oesophagus and proventriculus.

The beginning of the mid-gut or true stomach, as we regard it, is lined internally with a layer of large, long, erect papillae which extends nearly as far as opposite the end of the strainer, and is also extended along the outside of this organ. Just before a point opposite the end of the strainer, this layer of dense close-set papillae suddenly stops, and is succeeded by a division of the digestive canal lying between the point opposite the end of the proventricular strainer, and a point situated a little before the opening of the first pair of biliary ducts. This region, which we regard as the true stomach, has the inner surface raised into about twelve transverse or circular folds. Just where this region of the digestive canal ends, it contracts, and this is judged to be the line of demarcation between the mid-gut and hind-gut, i. e., the true stomach and the intestine. It should be observed that the chitinous folds of the oesophagus and proventricle (usually called stomach) are continuous, alike morphologically, and stop at the posterior end of the proventricular strainer. It is evident that the food, such as worms, at first partly torn by the teeth at the base of the limbs, is further triturated by the numerous hard teeth of the crop, while the more nutritious fluid portions strain through the narrow passage of the singular hollow cone.

The inner walls of the stomach are destitute of chitin, the long, close-set, large papillae being edged with a thick layer of columnar epithelium. The twelve circular folds of pavement epithelium are also lined with a similar columnar epithelium.

The four biliary ducts open into the intestine proper, which is lined as far as the rectal folds with an epithelial membrane, is divided by longitudinal and transverse lines into squares forming close-set, square, flattened papillae; on the posterior half of the intestine the longitudinal lines are more numerous than the transverse, the latter being partially obsolete, so that the inner surface of the intestine is gathered into fine longitudinal folds, the free edges of the folds being irregularly serrated.

These folds consist of pavement epithelium (or mucous membrane), the free edges of which are of columnar epithelium, the cells being long and narrow, while the nuclei are not so large and distinct as in the proventricle.

The intestine within suddenly contracts at the beginning of the rectum, but becomes larger posteriorly to the vent; the interior is thrown up suddenly into ten large folds of unequal size, which become smaller posteriorly. These rectal folds have the same muscular and epithelial layers as in the other parts of the digestive tract, but the cells of the pavement epithelium, instead of being uniformly round, are in places irregularly diamond or spindle-shaped, as in plate 5, fig. 6. The columnar epithelium of the rectal folds is lined externally (in the rectal cavity or lumen) with a lining of a clear, structureless, somewhat chitinous membrane which stains purple with haematoxylin. It would thus appear that the secreting surface of the stomach-walls is, owing to these folds and large erect papillae, very much greater than in the intestine. We have seen that the stomach, like the intestine, lacks the chitinous lining, and this, together with the histological identity of structure between what we regard as the stomach and the intestine, may seem to some as opposed to the view that this region is the mid-gut, stomach or archenteron; but the fact that it is divided from the intestine by a slight constriction, that it lies in front of the biliary ducts, and that the appearance and gross anatomy of the lining is unlike that of the intestine, coupled with the perfect continuity of structure in the oesophagus and proventricle, are to our mind sufficient arguments for the position we hold. Moreover in the lobster the two capacious biliary ducts empty directly into the true stomach or mid-gut, the small straight intestine beginning at some distance behind the origin of these ducts.

Thus while the stomach and intestine of Limulus agree in the absence of the chitinous layer, the rectum in its longitudinal folds and lining of chitine repeats in a degree the structure of the oesophagus.

Comparing the digestive canal of Limulus with that of the lobster or Decapodous Crustacea in general, we find that the oesophagus and so-called stomach (what we call in this paper crop or proventriculus), are continuous parts; that the true mid-gut or stomach has, like the intestine, no chitinous lining, though the rectum of the lobster, as we find on examination, has long rectal folds (besides large square raised projections), and is throughout lined with chitine. There is thus a general correspondence or homology between the Decapodous and Merostomatous digestive or enteric canal. Unfortunately we have been unable as yet to find any specimens of the young with the enteric canal in such an early stage of development as to throw any light on the morphology of the stomach.[1]

Structure of the liver. The tubules of the liver spread everywhere through the cephalothorax. reaching almost to the edge of the retina of the eyes. and when cut through, show in sections. as at plate 3. figs. 9, 9a, 9b. a circular or oval layer of epithelium, surrounding a cavity more or less irregular in size and form. The cells are quite large and filled with brownish granules, being dark at base and transparent towards the end where they project into the cavity.

Plate 3, fig. 8, represents the end of a lobule from the living horse-shoe crab. Compared with that of the lobster (plate 3, fig. 10), they are from one half to a third smaller, very much longer, more intestiniform, and contracted irregularly, while the pigment granules are thicker, and the entire mass is blackish-brown. Figs. 8a, 8b, 8c, represent the cells comprising the epithelium teased out and spherical in form. Fig. 8a, indicates a cell containing smaller nucleated cells of two kinds. the smaller clear and yellow, the larger, darker and horn-colored; 8b, a clear, nucleated cell; 8c. represents dark. clear amber-colored cells filled with the secretion, and with the nucleus no darker than the rest of the cell, and very clear. For purposes of comparison we give figures (plate 3, fig. 10) of the end of a liver-lobule of the lobster. which is pale green, with numerous epithelial cells, a few oil globules being scattered through them. In the living lobules of a species of Panopaeus common in Buzzard's Bay, some of the cells are colored yellowish-green. imparting the same color to the entire lobule; the cells in Panopaeus (plate 3, fig. 8d) are clear of granules, almost as much so as the fat globules. The lobules of the liver of this crab are larger, more conical and shorter than in the lobster. From this it will be seen that fundamentally both the general and minute structure of the liver of the Decapoda and Limulus is nearly identical.

The glandular bodies supposed to be renal in their nature. These glands had remained undescribed, until in a paper read at the Philadelphia meeting of the National Academy of Sciences, held in November, 1874.[2] we drew attention to their occurrence and histological structure. Although we have nothing to add verbally to the account then given of the gross anatomy of these glands, we would refer to the figure (plate 3, fig. 7) illustrating the form, and the cells (plate 3, figs. 7a, 7b, 7e) composing these glands. They do not appear to have been described by Van der Hoeven, Owen, or A. Milne-Edwards, in their account of dissections of this animal.

These glands are quite large. and apparently of some physiological importance, and are easily found, as they are conspicuous from their bright red color, causing them to contrast decidedly with the dark masses of the liver, and the yellowish ovary or greenish testes, near which they are situated. The glands are bilaterally symmetrical, one situated on each side of the proventricle and stomach, and each is entirely separate from its fellow.   Each gland (plate 3, fig. 7) consists of a stolon-

---

[1] I have found in the crop (stomach) of a large Limulus several *living* spiny larvae of *Homalomyia*, and several dead *Tottennia gemma*, mixed with bits of sea weed and Zostera. In another, occurred an Edwardsia, still alive, and three or four large *Nereis virens*.

[2] American Naturalist, ix, 511. September, 1875.

like mass (*a*), extending along close to the great collective vein, and attached to it by irregular bands of connective tissue, which also hold the gland in place. From this horizontal mass, four vertical branches (*b*, *b*) arise, and lie between and next to the partitions at the base of the legs, which divide the latero-sternal region of the cephalothorax into compartments. The posterior of these four vertical lobes accompanies the middle hepatic vein from its origin from the great collective vein, and is sent off opposite the insertion of the fifth pair of feet. Half-way between the origin of the vein and the articulation of the limb to the body, it turns at a right angle, the ends of the two other lobes passing a little beyond it, and ends in a blind sac, less vertical than the others, slightly ascending at the end, which lies just above the insertion of the second pair of feet. The two middle lobes are directed to the collective vein. Each lobe is somewhat flattened out, and lies close to the posterior wall of the compartment in which it is situated, as if wedged in between the wall and the muscles between it and the anterior portion of the compartment. Each lobe also accompanies the bases of the first four tegumentary nerves.

I could not by injection of the gland discover any general opening into the coelom or body cavity, or perceive any connection with the hepatic, or with the great collective vein. The four lobes end in blind sacs, and have no lumen or central cavity.

The lobes are irregular in form, appearing as if twisted and knotted, and with sheets and bands of connective tissue enclosing the muscles, among which the gland lies. Each lobe when cut across, is oval, with a yellowish interior and a small central cavity.

The gland externally is of a bright brick-red. The mass is quite dense, though yielding, and on this account hard to be cut with the microtome.

When examined under Hartnack's No. 9 immersion lens and Zentmayer's B eye-piece, the reddish external cortical portion when teased out from the living animal, is seen to consist of closely aggregated, irregularly rounded, nucleated cells of quite unequal size; and scattered about in the interstices between the cells, are dark reddish pigment masses (plate 3, fig. 7*a*) which give color to the gland. They are very irregular in size and form, and twenty hours after a portion of the living gland was submitted to microscopic examination moved to and fro. In other portions of the outer reddish part of the gland, where the pigment masses are wanting, the mass is made up of fine granular cells, which have no nucleus. Other cells have a large nucleus filled with granules, and containing nucleoli.

In the yellowish or medullary portion are scattered about very sparingly certain spherical cells which probably are purely secretory (plate 3, figs. 7*b*, 7*c*). The nucleus is very large and amber colored, with a clear nucleolus; others have no nucleolus, and the small ones are colorless.

I am at a loss to think what these glands, with their active secreting cells filled with a yellowish fluid, can be, unless they are renal and excretory in their nature. In general position they coincide with that of the shell glands of the Entomostraca and Branchiopoda, including the Phyllopoda, especially as worked out in Leptodora, by Weismann.[1] But in lacking apparently an excretory duct, and in their dense parenchym, with no lumen, as well as histologically, they seem to differ from the shell glands of the lower

[1] Ueber den Bau und Lebenserscheinungen von Leptodora hyalina. Zeitschr. für wiss. Zoologie, Bd. xxiv. p. 385, 1874.

Crustacea, and the green glands of the Decapoda. It should be remembered, nevertheless, how difficult it is to find the excretory duct of the green gland, though its inlet is very apparent. It is probable that we have here to deal with a new form of kidney, adding a fourth kind to the three forms of renal organs existing in the Crustacea.[1]

STRUCTURE OF THE EYES OF LIMULUS.

After we had made some researches on the structure of the compound eyes of Limulus, and had ascertained that their structure is quite unlike that of other Arthropodous eyes, having a chitinous lens and no rods and cones, we had the opportunity of examining Grenacher's elaborate work entitled *Untersuchungen über das Schorgan der Arthropoden, insbesonderer der Spinnen, Insecten und Crustaceen.*[2]

We have little to add to Grenacher's account of the histology of the compound eye, and our studies confirm the accuracy of his account and the three drawings he publishes of the structure of the retinula and the rhabdom, although we have failed to find the layer of epithelial cells extending up between the corneal lenses and next to the pigment layer; these are much less regular in their arrangement than drawn by Grenacher, and seem to be simply connective tissue cells, which are as abundant away from the lenses as next to them. We have also not succeeded in observing that the optic nerve fibres end in the manner indicated in his drawing. We may here say that we had examined sections of the compound eye of Limulus, and had made out the leading points in its structure before seeing Grenacher's work.

The following account is based upon observations made upon sections cut for us by Mr. Mason. They are taken in most cases from living specimens, placed in alcohol, and hardened in gum arabic; and either stained with picro-carmine, or else the pigment layer dissolved wholly or in part with nitric acid in order to uncover the ends of the corneal lenses and to show the structure of the retinula and rhabdom. The subdivisions of the optic nerve were best showed in slices stained with picro-carmine, the nervous substance being but partially colored and contrasting well with the highly tinged connective tissue by which the nerves are surrounded. In order to study the eye of Limulus intelligently Mr. Mason kindly made for us numerous exquisite sections of the eye of the lobster.[3]

Plate 6, fig. 1, represents an actual section of the eye, with its exterior convex surface, its lenses, retina and nerves. The surface of the eye is convex, smooth and polished. The integument over the eye suddenly diminishes in thickness to form the cornea; it is solid and chitinous as in the rest of the integument, and is composed of three layers: the outer and thinner more solid one (1_1), which is clear yellow or amber-colored; the middle (1_2), which is duller yellow and is finely laminated and softer, being partially

[1] See Eng. Wassiljew. Ueber die Nieren des Flusskrebses. Zoologischer Anzeiger, p. 221, 1878.

[2] Von H. Grenacher. Göttingen, 1879, 11 lith. taf. 40, pp. 188.

[3] These sections made by Mr. Mason were remarkably successful, the slices being thin enough to include a layer of hundreds of facets and rods and cones but one deep, with the optic nerve-fibres, so that the structure of the eye could readily be studied. We did not perceive that the anatomy of the eye of *Homarus americanus* differed in any important respect from that of the European lobster as worked out by Mr. Edwin T. Newton. Quart. Journ. of Microscopical Science, 1873, p. 339.

stained reddish by carmine; and a third (l_3) thicker layer, less laminated and pierced by nutritive canals (p), filled with connective tissue and directly communicating with the body cavity. From the cornea project obliquely inwards large, long, solid, conical processes (cl). These are the "corneal lenses" of Grenacher, which he regards as homologous with the corneal lenses of larval insects and of Arachnida. We see no reason to dissent from this opinion. These corneal lenses are long, cylindrical, obtusely pointed, sometimes quite sharp, at the end. They point inwards more or less obliquely towards the centre of the eye. Those (as at fig. 2a) near the periphery of the eye are longer and slenderer and more oblique than those in the centre, the latter being considerably shorter and blunter (plate 6, fig. 2). These lenses are developed from the third, a portion of the second or more laminated layer of the cornea filling up a conical space (fig. 2a, h,) at the base of the cone; the laminae composing this shallow cone within the larger cone are continuous with the laminated layer of the cornea, and like it are stained reddish by the carmine, while the cone itself remains unstained, of a clear amber color, and is structureless; sometimes one or two curved lines being seen parallel to the periphery of the end of the cone. That the corneal lens is solid is proved not only by its appearance, as seen in numerous sections, but by the frequent marks of the razor, and by the laminated structure of the inner conical portion. What relation, if any, the conical part (h) has physiologically to the corneal lens, we are not prepared to state.

The terminal half, or sometimes third, of the corneal lens is enveloped in the pigment layer or retina, (plate 6, fig. 3, rt), which is morphologically a continuation or modification of the dark hypodermis (hy). The layer is continuous between the ends of the solid corneal lenses, but is produced at the ends of the latter into cones of corresponding size (rtc), which project into the body-cavity, and are enveloped by the dense connective tissue. As stated by Grenacher, this pigment layer is composed of modified epithelial cells, which are very long and slender, with a minute nucleus (fig. 3, rcl). It is very difficult to make out these cells, and we should have overlooked them had not Grenacher described and figured them; finally, however, we could trace them, in preparations treated with acid, into the hypodermis, where the cells are also long and slender, though shorter than in the retina. Plate 6, fig. 3, rcl, represents these retinal cells, as seen at the end of an acute corneal lens, and their relation to the rhabdom ($rhab$).

Besides the retina, the soft parts of the entire compound eye of Limulus consists of a large mass of connective tissue (ct), lying under and next to the retina and finely granular, permeated by the irregular tortuous branches of the optic nerve. The cells and granules of this specialized subocular portion of the connective tissue forming the parenchyma of the cephalothorax are smaller than elsewhere; they are nucleated, and the tissue stains paler crimson by the picro-carmine, than the connective tissue beyond the subocular area, which remains darker brown, with coarser granules. The arteries, ovarian-tubes and liver-tubes, rarely penetrate into the subocular area; and the branches of the optic nerve do not wander into the region beyond. Fig. 1, ar, represents the cut ends of two minute arterial branches, ov represents the cell-eggs of the end of an ovarian tube, and l indicates the much larger sections of a liver-tube; these vessels constitute the greater part of the soft portions of the cephalothorax, being brown or yellowish brown, and enveloped in a dense connective tissue.

The subdivisions of the optic nerve can rarely be traced for a long distance continuously, owing to their irregular, tortuous course. In the drawing (fig. 1), I have delineated with the aid of the camera lucida an actual section; the clear spaces indicate the cut portions of the nerves distributed to each corneal lens. Histologically they present the same appearance as the nerve-fibres in the brain, those given off from the lower ganglionic cells. Under a low power (½ inch), they appear to be structureless; under a ¼ they are seen to be finely granulated. After repeated search I could find no nuclei in these fibres; nor were there any ganglion cells to be discovered. Repeated examinations of numerous sections treated in different ways, have convinced me that there are throughout the subocular area no ganglion cells, such as are characteristic of the eyes of spiders and myriopods; hence, with Grenacher, we may state that a *ganglion opticum* is entirely wanting in Limulus; the irregular, tortuous subdivisions of the optic nerve are sent directly to the corneal lenses.

Coming now to the structure of the individual eye, or facet in the compound eye, we find that its anatomy is just as described by Grenacher, except that we have been led to doubt the existence of the layer of triangular (in outline) cells, which he represents as extending up between the conical corneal lenses, and it should be borne in mind that we examined eyes taken directly from living Limuli, as well as specimens that had been preserved in alcohol for several years. Grenacher's researches were made on eyes preserved for a long time in specimens of Limuli obtained from German museums, and his material was so poor that he did not attempt to study the simple eyes (ocelli).

The structure of the cone of pigment matter enclosing and extending beyond the end of the conical line has been described and illustrated in a masterly manner by Grenacher. Impinging on the end of the conical lens, and extending through the centre of the conical mass is a twelve-radiate semi-solid body, called by Grenacher the rhabdom, and which he apparently homologises with the rhabdom or spindle-shaped body, succeeding the rod of the ordinary Crustacean eye. Along a part of its length, this rhabdom (fig. 4, *rhab*) is enveloped by the retinula (plate 6, fig. 4, *ret*). Our figures show in sections the rhabdom, with its central axis and twelve or thirteen rays, forming a rosette extending into the substance of the retinula. That the retinula is, as Grenacher figures, composed of as many large cells as there are rays of the rhabdom, we have proved by preparations treated with acid, as seen in fig. 4, *a*.

How the optic nerve is connected with or impinges on the rhabdom, we have been unable to ascertain. We have only seen enough to convince us that the nerve reaches the end of the rhabdom, but the nature of the ending is unknown to us. The nerves, as seen in our drawing, fig. 1, sometimes appear as if they ran by the end of the retinal cones, and extended up between the corneal lenses. On the other hand, we have seen very plainly the mode of termination of the nerve in the ocellus. Grenacher, however, states that "a number of the nerve-fibres are distributed to each single-eye [facet], they diverge behind it, and I have repeatedly traced clearly the entrance of a fibre into the axial part of a retinula cell."

Grenacher concludes that perception in the typical Arthropod eye is performed according to the mosaic theory of Müller, and that this applies to the eye of Limulus, although the eye of the latter is morphologically wholly different from the eyes of any other animal.

According to Grenacher, the conical lenses are not homologues of the crystalline lenses of other Arthropods. and the eye of Limulus cannot, he holds, be compared with the eyes of any other Arthropoda. There can be, he claims, no genetic connection between the eye of Limulus and those of any other Arthropods, and the two types of eye, *i. e.,* those of Limulus and all other Arthropods, agree only in the fact that they are compound. Among the Arachnida, he states, one may seek in vain for such an isolated type of eye. He adds : " But it is not only possible but also probable, that the Poecilopoda are related by their eyes to Myriopoda. In Cermatia, the eyes are wholly unlike those of the spiders or insects, and they seem to have something in common with those of Limulus." We shall see further on, however, that the type of eye of Cermatia is not fundamentally unlike that of Bothropolys, and other Myriopoda, as figured by Graber.

We have seen, then, that there is in the eye of Limulus an entire absence of rods and cones, a common feature of the Arthropod eye. The corneal lens of Limulus corresponds to the cornea or facet of each individual Arthropod eye, but there are no rods and cones, no optic ganglion, no scattered ganglionic cells, but the end of the long, solid, conical, corneal lens is simply enveloped by the pigment mass, and the end of the cone is succeeded by a rhabdom, partly enveloped by the retinula, the terminus of the optic nerve passing into the axial part of a retinula cell.

Comparison of the compound eye of Limulus with that of Trilobites. Beyond the fact that the entire eye of certain Trilobites, and enlarged views of the outer surface of the cornea of the eye, have been described and figured in Burmeister's work on the organization of Trilobites and in various palaeontological treatises in Europe and North America, especially by Barrande in his great work on Trilobites, I am not aware that any one has given a description of the internal structure of the hard parts of the eye of Trilobites.

The full bibliography of treatises relating to these animals in Bronn's Die Classen und Ordnungen des Thierreichs, carried up to 1879 by Gerstäcker, contains references to no special paper on this subject, and the résumé by Gerstäcker of what is known of the structure of the eye, only refers to the external anatomy of the cornea, the form of the facets and their number in different forms of Trilobites. He shows that observers divide them into simple and compound ; the former (ocelli) are found in the genus Harpes. These " ocelli" are said to be situated near one another, and are so large that the group formed by them can be seen with the unaided eye ; the surface of the single " ocellus" appears, under the glass, smooth and shining. From the description and the figure of the eye enlarged, from Barrande, it would seem as if each eye was composed of three large simple ones ; so that these eyes are really aggregate, and not comparable with the simple eye or ocellus of Limulus and the fossil Merostomata.[1] Moreover, the situation of these so-called ocelli is the same as that of the compound eyes of other Trilobites.

The Trilobites with compound eyes are divided into two numerically very dissimilar groups ; the first comprising Phacops and Dalmanites alone, and the second embracing

---

[1] The eyes of the fossil Merostomata (Eurypterus and Pterygotus) are evidently in external form and position, judging by Mr. Woodward's figure, exactly homologous with the ocelli and compound eyes of Limulus.

all the remaining Trilobites, excepting of course the eyeless genera, Agnostus, Dindymene, Ampyx and Dioride. The eyes of Phacops and Dalmanites are said by Quendstedt and Barrande not to be *compound* eyes in the truest sense, but *aggregated* eyes (*Oculi congregati*). But judging by Barrande's figures of the eyes of *Phacops fecundus* and *P. modestus* (Barrande, Vol. I, Suppl. Plate 13, figs. 12 and 22), and our observations on the exterior of the eye of an undetermined species of *Phacops*, kindly sent us by Mr. J. F. Whiteaves, Palaeontologist of the Canadian Geological Survey, we do not see any essential difference between the form and arrangement of the corneal lenses of Phacops and Asaphus, and are disposed to believe that the distinctions pointed out by the above named authors are artificial.

For my material I am mainly indebted to Mr. C. D. Walcott, who has so satisfactorily demonstrated the presence in Trilobites of jointed cephalothoracic appendages. On applying to him for specimens, and informing him that I wished to have sections made of the eyes of Trilobites to compare with those of Limulus, he very generously sent me his own collection of sections of the eyes of *Asaphus gigas* and *Bathyurus longistrinosus*, which he had prepared for his own study, also other eyes, and especially the shell or carapace of a large Asaphus, from Trenton Falls, showing the eye and the projecting points of the corneal lenses. Prof. Samuel Calvin kindly sent me the eyes of an unknown Trilobite from the Trenton limestone, one specimen showing the pits made in the mud by the projecting ends of the corneal lenses, while to Mr. Whiteaves I am indebted for eyes of Calymene.

First turning our attention to the casts and natural sections; that of the interior of the carapace, including the molted cornea of *Asaphus gigas*, is noteworthy. When the concave or interior surface of this specimen is placed under a magnifying power of fifty diameters, the entire surface is seen to be rough with the ends of the minute solid conical corneal lenses which project into the body-cavity. This is exactly comparable with the cast shell of Limulus and its solid corneal lenses projecting into the body-cavity (plate 6, fig. 1). Those of Asaphus only differ in being much smaller and more numerous, and perhaps rather more blunt. Without much doubt the ends of the corneal lenses of Asaphus, as in Limulus, were enveloped in the retina, the animal molting its carapace, the hypodermis with the retina being retained by the Trilobite, while the corneal lenses were cast with the shell.

In the specimen of the unknown Trilobite from Iowa received from Prof. Calvin, the corneal lenses, seen externally, are quite far apart, arranged in quincunx order; the lenses are round and decidedly convex on the external surface. In a natural section, where the eye has been broken into two, the conical lenses are seen to extend through the cornea as cup-shaped or conical bodies, and are quite distinct from the cornea itself. In another broken eye of the same species, the cornea is partly preserved, and two of the corneal lenses are seen to extend down into and partially fill two hollows or pits; these pits are evidently the impressions made in the fine sediment which filled the interior of the molted eye or cornea!

Thus in the *Asaphus gigas* noticed above, we have the entire inside of the cornea with the cone-like lenses projecting from the concave interior; while in the last example we have the impressions made by the cones in the Silurian mud which silted into the cornea after the Trilobite had cast its shell.

Farther evidence that the Trilobite's eye was constructed on the same pattern as that of the living horse-shoe crab is seen in the sections made by Mr. Walcott. We will first describe, briefly, the eye of Limulus. Plate 6, fig. 1 represents a section through the cornea of Limulus; *cor*, the cornea which is seen to be a thinned portion of the integument; *pc*, indicates one of the nutrient or pore canals, which are filled with connective tissue extending into the integument from the body cavity; *cl*, is one of the series of solid conical corneal lenses. These are buried partly in the black retina, and the long slender optic nerve just before reaching the eye subdivides, sending a branch to each facet or cornea, impinging on the lens. Fig. 6 represents a vertical view of the corneal lenses or facets, magnified fifty diameters, as seen through the transparent cornea. It will be seen that they are slightly hexagonal and arranged in quincunx order; their external surface is flat, though that of the ocelli is slightly convex.

Now if we compare with the horse-shoe crab's eye that of the trilobite, *Asaphus gigas*, (plate 6, figs. 8, 9), we see that the eye is raised upon a tubercle-like elevation of the carapace; the integument (*int*) is about as thick as that of Limulus, and it contains similar pore-canals (*pc*); the eye itself or cornea, occupies a rather small area; its exterior surface, instead of being smooth as in Limulus, is tuberculated, or divided up into minute convex areas; these convexities are the external surfaces of the corneal lenses, which extend through the cornea, so that its surface is rough instead of smooth as in Limulus; *cl* indicates one of the corneal lenses which are arranged side by side; they are of slightly different lengths and thicknesses, and the rather blunt free ends project into the cavity of the eye, which in the fossil is filled with a translucent calcite.

It is quite apparent that we have here the closest possible homology between the hard parts of the eye of Limulus and of Asaphus. Another point of very considerable interest is a tolerably distinct dark line (fig. 9, *rt*), which seems to run across from one lens to another, and which may possibly represent the external limits of the retina or pigment mass in which the ends of the lenses were probably immersed; should this be found to be the indication of the outer edge of the retina, it would be a most interesting fact in favor of our view of the identity between the eyes of the two types of Palaeocarida under consideration.

Another section sent us by Mr. Walcott is represented by plate 6, fig. 8; it is from *Asaphus gigas*, but represents a less elevated and broader part of the eye than that seen in plate 6, fig. 9; the section does not so well exhibit the free ends of the corneal lenses. Fig. 7 *a* represents a tranverse view of the eye of *Asaphus gigas*, showing the hexagonal form of the facets, and their quincunx arrangement; so much like that of Limulus (fig. 6).

This hexagonal appearance of the corneal lenses is still retained in natural vertical sections of eyes of the same genus, where with a good Tolles lens the sides of the cones are seen to be angular. Plate 6, fig. 10, represents a few such cones. I do not understand to what this hexagonal appearance is due; for both in Limulus and the Trilobites the corneal lenses appear usually to be round, and yet in making a camera drawing (as are all those here represented) of the cornea of Limulus from above, they present the same hexagonal appearance as in Trilobites. The cause of this I leave to others to explain.

In a section (transverse) of the cornea of *Bathyurus longistrinosus*, received from Mr. Walcott, the lenses are seen to be very irregular, five- or six-sided, and very irregularly grouped, not arranged in distinct rows.

From the facts here presented, it would seem evident that the hard parts of the eye of the Trilobites and of Limulus are, throughout, identical. The nature of the soft parts will, as a matter of course, always remain problematical; unless the dark line indicated in plate 6, fig. 9 (*rt?*) really represents the outer edge of the pigment of the retina; but however this may be, judging by the identity in structure of the solid parts, we have, reasoning by analogy, good evidence that most probably the eye of the Trilobites had a retinal mass like that of Limulus, and that the numerous small branches of the long slender optic nerve (for such it must have been) impinged on the ends of the corneal lenses. It has been shown by Grenacher and myself that the eye of Limulus is constructed on a totally different plan from that of other Arthropods; I now feel authorized in claiming that the Trilobite's eye was organized on the same plan as that of Limulus; and thus when we add the close resemblance in the larval forms, in the general anatomy of the body-segments, and the fact demonstrated by Mr. Walcott that the Trilobites had jointed round limbs (and probably membranous ones), we are led to believe that the two groups of Merostomata and Trilobites are subdivisions or orders of one and the same subclass of Crustacea, for which we have previously proposed the term Palaeocarida.

Structure of the simple eyes or ocelli. Owing to insufficient material, Grenacher did not study the simple eye of Limulus. The structure of an ocellus repeats very closely that of one of the individual facets or members of the compound eye. At the point where a simple eye is situated, i. e., on each side of the median spine near the front edge of the carapace, the chitinous integument suddenly becomes much thinner; the integument is divided as in the cornea of the compound eye into three portions, an outer thin yellow clear portion; a much thicker finely laminated part with fine granules and capable of being stained reddish; and a much thicker clear part, which has about a dozen layers, not seen in the third inner layer of the integument next to the edge of the compound eye. The integument is also penetrated throughout by canals filled with connective tissue. The surface of the cornea is slightly convex. Next to the base of the large corneal lens, there is a chitinous portion (*p*) which is less dense than the adjoining clear part, becoming stained a pale crimson by picro-carmine. Just as in the facets of the compound eye the laminated part of the cornea extends conically into the base of the corneal lens, forming a cone within the larger lens; this part (*h*) is less dense than the lens, and is usually more distinctly conical than the lens itself. The latter is a large solid mass of chitin, with two curved lines (plate 5, fig. 13, *cl*) in some examples, showing a slight tendency to lamination; in form it is longer than thick, and very obtusely rounded at the end, being as thick near the end as at the base; in form therefore the lens differs decidedly from that of the corneal lenses of the compound eye. That the corneal lenses of both simple and compound eyes of Limulus are solid is proved by the fact that they do not stain reddish like the laminated portion of the cornea and adjacent integument, and also because they are excavated as solid cones projecting inwards from the cast chitinous crust of the animal. In a specimen (fig. 12), not treated with acid, the end of the cone is seen to be buried in pigment, and in one out of many sections, i. e., that figured 14, the cut went directly through the ocellar nerve, which, as seen in the figure, after leaving the branch distributed to the other ocellus, proceeds undivided to a distance about equal to the diameter of the corneal lens, when it gives off minute fibres which pass up and lose themselves in the pigment layer near the base of the cone. The main nervous trunk is seen to impinge directly on the end of the pigment mass surrounding the end of the lens, while branches pass up into the pigment on each side of the lens; so that the latter is immersed, so to speak, in a multitude of nervous fibres.

On treating the pigment with acid, and cutting a section on one side of the solid lens, as at plate 6, fig. 5, the entire mass of connective tissue and pigment is seen to be permeated with nerve-fibres, which end in slight, bulbous, partly hyaline expansions next to the chitinous integument.

Nothing like the rhabdom or retinula was to be observed, and I doubt much if they exist, or any nucleated ganglionic cells.

We have, then, in the simple eye or ocellus of Limulus a repetition of the general structure of any one of the individual eyes of the compound organ of vision, without the rhabdom and retinula. The simple eye, then, in the horse-shoe crab is apparently rather more rudimentary than one of the elements of the compound eye; and it is difficult to conceive of a much simpler form of eye in an arthropodous animal; hence it can not be said of the ocellus of Limulus that it is not less primitive in structure than the compound eye; for we have here the eye reduced to a corneal lens, retina and optic nerve, the simplest association of elements in any organ of vision.

Comparison of the ocellus of Limulus with the eyes of Myriopods. When we compare the ocellus of Limulus with that of the Arachnida, and of larval insects, there is very considerable difference. In the form of the corneal lens, however, the ocellus of Limulus somewhat approaches that of the Myriopods, as lately worked out by Graber.[1]

An examination of the agglomerated eye[2] of *Bothropolys bipunctatus* Wood — a genus allied to Lithobius, the species here named being common in Northern California at the base of Mount Shasta — shows us that the myriopod eye, as a whole, is entirely unlike that of Limulus.

The brain, in the first place, is on the usual Arthropod type; the hemispheres being symmetrical, and the relative position of the larger ganglion-cells being like those of hexapod insects. A large mass of ganglion-cells is situated at the origin of each optic nerve. As regards the eye or group of eyes, the individual eyes are about eighteen in number and closely aggregated, though each simple eye or facet is circular and its surface convex. The cornea in the specimen examined, while externally convex, is concave on the inside, the cornea being no thicker in the middle than

[1] Ueber das unicoreale Tracheaten- und speciell das Arachnoiden- und Myriopoden-Auge. Von V. Graber. Archiv. für mikr. Anat. Bd. 17, heft. 1. Aug. 14, 1879. [2] Our sections were kindly made for us by Mr. Mason.

on the sides. not being lens-shaped as usual in the Myriopod and Arachnidan eyes, as described by Graber and Grenacher. The cornea is laminated, as in the integument. That the cornea is apparently normally concave in this genus, seems evident from the fact that the soft parts next to, be described fill the concavity of the cornea. The solid parts, then, of this Myriopod, are quite unlike the larger corneal cones of Limulus; though in general, the corneal lens of the Myriopods examined by Graber appear to be homologous with the cones of Limulus.

When we compare the soft parts of the eye of *Bothropolys* and the Myriopods in general with those of Limulus, we find nothing in common.

In Bothropolys the soft parts consist of the layer of rather large, round, nucleated, epithelial cells, situated next to the cornea, and called by Graber the " lens-epithelium or vitreous-body cells." (Glaskörperzellen). This layer (absent in Limulus) is succeeded by the layer of short, slender-pointed rods, as figured by Graber, with large nucleated cells in the tissue enveloping them. This layer of rods, homologues of the rods of other Tracheate and Crustacean eyes (also absent in Limulus) is succeeded by the retina, a continuation of the hypodermal epithelial layer, the cells of the latter being much more distinct and larger than the hypodermal cells of Limulus.

The retina, whose structure differs from that of Limulus in having no "retinula," is succeeded by the *ganglion opticum*, (absent in Limulus), which consists of a layer of very large ganglion-cells rounded and overlapping each other; their fibres leading away from the eye, to form the optic nerve.

Eye of Cermatia forceps. The eye of this Myriopod appears to be constructed on the same plan as that of other myriopods though differing in some important respects. Though Cermatia is said to have compound eyes in contradistinction from the ocelli of other myriopods, the latter are truly aggregated or compound, the so-called " ocelli" being made up of contiguous facets, the nerve fibres which supply them arising in the same general manner from the optic nerves.

The following description is made from sections made by Mr. N. N. Mason of Providence, and loaned me for description : —

The eye is composed of a hemispherical, many facetted cornea, the lenses of which are shallow, doubly convex, being quite regularly lenticular, the chitinous substance being laminated as usual.

Each corneal lens is underlaid by a retina about as thick as the cornea, the inner surface of each retinal mass being convex. Corresponding to each lens is a separate mass of connective tissue, which increases in thickness from the end of the optic nerve outwards towards the cornea, there being usually a clear interspace between each mass. Within the broad stratum of connective tissue, next to the corneal lens within the retina, is a layer of rounded " vitreous cells" or " or lens-epithelium" of Graber. This layer is succeeded by the series of rather large visual rods, one in each mass corresponding to each corneal lens; these rods are long and sharp, conical at the end, the ends extending one-half to two-thirds of the distance inwards to the inner edge of the retinal mass; they each possess a nucleus, and the connective tissue enveloping the rods is nucleated, while there is an irregular layer of nucleated cells near or around the ends of the rods.

This layer of cells is succeeded by a thin, slightly curvilinear, transverse strip of connective tissue, passing through the entire eye, and behind it are the loose, nucleated spherical cells forming the *ganglion opticum*, among which the fibres of the optic nerve pass.

The brain of *Cermatia forceps*, as shown by several sections, is modelled on the same plan as we have observed in Bothropolys and so far as we see, the myriopod brain corresponds more closely in its general form and structure with that of the Insects than of the Crustacea. The large, thick optic nerve arises from the upper side of each hemisphere. The median furrow above is deep, and on each side is a mass of small ganglion-cells; also a mass in the deep sinus below the origin of the optic nerve, and another mass on the inferior lobe extending down each side of the oesophagus, probably near or at the origin of the posterior commissure. These masses, *i. e.*, those on the upper and under side of the brain, connect on each side of the median line, and in this respect the brain is as in Bothropolys. There are no large ganglion cells, as in the Crustacea and in Limulus.

It will be seen from this brief account that the eyes of Limulus differ from those of Myriopods in wanting the lens-epithelium, the rods, and a *ganglion opticum*, which are present in other Arthropods, both tracheate and branchiate.

The Brain and its Internal Structure.

Several years ago, before the present interest in the study of the brain of Arthropoda had arisen, I made an attempt to study the brain or supra-oesophageal ganglion of the horse-shoe or king crab. Mr. T. D. Biscoe kindly cut a number of sections for me. These were unstained, and owing to interruptions were not examined until the past winter, when with the aid of a large number of other sections made by Mr. N. N. Mason of Providence, R. I., I have been enabled to present the following results. The brain was in some cases stained with osmic acid in the manner described by Dietl [1] and adopted by Newton — being taken from the living animal and allowed to remain in the osmic acid from twenty to forty hours.

The best results, however, were obtained from sections of two brains, one of which had been several years in alcohol, and which took the osmic acid stains very evenly; better results ensued by staining, after hardening the brain for two or three days in alcohol; the brain is so large that it does not harden readily, when put fresh and living in osmic acid alone.

When sections were not properly stained in the centre with osmic acid, they were further treated with a picro-carmine stain with good results. Mr. Mason embedded the brain, when

[1] The following articles bearing on the brain of the Arthropods have been consulted; the actual bibliography of the subject being somewhat fuller.

Ofsiannikoff. Ueber die feinerer Structur des Kopfganglions bei den Krebsen, besonders beim Palinurus locusta. Von Ph. Ofsiannikof. Mém. Acad. Imp. Sc. St. Petersbourg. Tom. VI. No. 10, 1863.

Dietl. Die Organization des Arthropodengehirns. Von M. J. Dietl. Zeitschr. wissensch. Zool. Bd. 27, 1876, p. 488.

Flögel. Ueber den einheitlichen Bau des Gehirns in den verschiedenen Insectenordnungen. Zeitschr. wissensch. Zoologie, Bd. xxx, Suppl. 1878, p. 556.

Newton. On the brain of the Cockroach, Blatta orientalis. By E. T. Newton. Quart. Journ. Microscopical Science, July, 1879, p. 340.

Krieger. Ueber das Centralnervensystem des Flusskrebses. Von K. R. Krieger. Zeitschrift für wissenschaft. Zoologie. Bd. xxxiii, Jan. 23, 1880, p. 527.

hardened and after remaining twenty-four hours in gum water, in a mixture of equal parts of paraffine, wax and olive oil, so that the consistency of the imbedding substance was nearly as soft as the tissue to be cut. The sections were made by a microtome devised by Mr. Mason, and were mounted in glycerine jelly.

While between two hundred and three hundred sections were made, by far the best results were obtained by a series of fifty-six sections, cut by Mr. Mason from one brain, and forty-four from the upper four-fifths of another brain, the slices being either $\frac{1}{1000}$ or $\frac{1}{500}$ of an inch in thickness; the best results of course were obtained from the thinner sections. These were deeply stained of a dark-brown, the ganglion-cells and nerve-fibres being much lighter than the nucleogenous masses forming the larger part of the brain, these being dark brown, the former tawny or yellowish brown.

The examination of a few sections of the brain of the lobster and the locust also kindly prepared by Mr. Mason, enabled me the more readily to understand the recent papers of Dietl, Newton and Krieger on the brain of the crawfish and insects, and afforded a standard of comparison with which to study the topography and histology of the brain of Limulus.

General anatomy of the brain. The position of the brain in relation to the body walls and digestive tract is seen in the section of the adult animal on plate 2, fig. 1, *br.* The central nervous system consists of an oesophageal collar made up by the consolidation of six pairs of postoral ganglia from which nerves are distributed to the six pairs of gnathopods. The ring is closed in front by the supra-oesophageal ganglion, or the partial homologue of that pair of brain-centres in the normal Crustacea and Insects. It will be remembered that in these Arthropods the brain is situated in the upper part of the head, in a plane parallel to, but quite removed from, that of the rest of the ganglionic chain; in Limulus, however, the brain is situated directly in front of and on the same plane with the horizontal oesophageal collar, and the abdominal portion of the central nervous system.

We now come to the singular relations of the ventral system of arterial vessels to the nervous system. This is fully described by A. Milne-Edwards. After describing the vascular ring surrounding the oesophagus he remarks : " Lorsqu'on ouvre cette portion du système artériel, on trouve dans son intérieur le collier nerveux oesophagien, le reste de la chaine ganglionnaire et la plupart des principaux nerfs qui y sont baignés par le sang. Les artères ne sont par seulement appliquées sur le système nerveux, comme chez les scorpions, ou développées à la surface de ce système de façon à le recouvrir ; elles logent celui-ci dans leur cavité. Cette disposition rappelle celle du réservoir sanguin dans l'intérieur duquel M. de Quatrefages a constaté l'existence des ganglions cérébroïdes chez les Planaires, et celle du vaisseau ventral des Sangsues, découvert par Johnson." He then states that these singular relations of the apparatus of innervation with the arterial system of Limulus have been seen, but very incompletely, by Professor Owen, and are more intimate than this eminent anatomist thinks, and quotes as follows from Owen's Comparative Anatomy and Physiology of Invertebrate Animals (1855, p. 320) : " The two large lateral branches (celles qui j'appelle les crosses aortiques) form arches which curve down the side of the stomach and the oesophagus, giving branches to both those parts and to the intestine, and becoming intimately united with the neurilemma of the oesophageal nervous collar. They unite at the posterior part of that collar, and *form a single vessel, which accompanies the abdominal nervous ganglionic chord* to its posterior bifurcation, when the vessel again divides. *Throughout all this course, the arterial is so closely connected with the nervous system as to be scarcely separable or distinguishable from it. The branches of the arterial or nervous trunks which accompany each other may be defined and studied apart.*" Afterwards in his "Anatomy of the King Crab," p. 24, Professor Owen thus writes of the arterial system: " On each side the origin of the ' ocellar artery arises one of double the size (*ib., e.e*), which. diverging from its fellow, curves outward and downward over the fore-part of the intestinal canal (plate 2, A. fig. 1 s); it gives off, in this course, a branch which ramifies upon the gizzard, a second to the intestine and liver, the main trunk being continued to the nervous annular centre where it expands, and combines with its fellow of the opposite side to form a sheath for that centre analagous to a 'duramater.' This rather loose sheath is continued along the ganglionic ventral cord, and is prolonged, like a loose neurilemma, upon the nerves sent off therefrom, as it is upon those in connection with the annular centre." [1]

Our own dissections and microscopic sections have taught us that the brain is enclosed by a thick neurilemma, which is different histologically from the arteries, containing no muscular layer. This layer closely envelops the brain-substance, and there is certainly no space between the brain and its neurilemma for the passage of the blood. Now the two lateral arteries descend from the anterior end of the heart, and open just behind the brain into the space between the oeosophagal ring, and its neurilemma, so that the latter is bathed in blood ; the artery merges into the neurilemma, the sinus being largest on the upper side of the oesophageal ring. On each side of the back of the brain is a large artery for the supply of the brain. but there are no small arterial branches. The whole nervous cord behind the brain, including the ganglionic enlargements, is loosely invested by this neurilemma, the space being very wide between the nervous cord and its loose coat, so that the nervous cord and ganglia are directly bathed by the blood. This neurilemma (or perineurium) also invests the larger nerves sent off from the ganglia. That the nervous cord fills up but a portion of the space within this outer coat may be seen by reference to plate 6, figs. 12–14. In embedding the portions of the nervous cord to be cut, the interspace is filled with the paraffine preparations. We thus conclude, that while Owen and Milne-Edwards' view is substantially correct, it should be modified somewhat, viz., the blood does not flow around the brain itself, though it may flow around the nerves sent to the simple and compound eyes ; and the nervous system appears to us not to be surrounded by a true artery, but that the thick perineurium becomes a vicarious arterial coat.

The brain in a Limulus ten inches long, exclusive of the caudal spine. is about six millimetres in diameter; it is broad and flat above, and on the under side full and

---

[1] Having only the first edition of Owen's Lectures on the Invertebrates in my library. I can not verify the quotation above made from the edition of 1855. In a recent letter from that distinguished anatomist he quotes as follows from p. 309. "The sides of the great oeosophageal ring are united by two transverse commissural bands; but the most remarkable feature of the nervous axis of this Crustacean is its envelopment by an arterial trunk." From this it would appear that Professor Owen was the first to perceive that the nervous cord is *enveloped* by the artery, though these organs were afterwards elaborated described and figured by A. Milne-Edwards.

rounded; on the upper side is a broad, shallow, median furrow, indicating that it is a double ganglion. Three pairs of nerves and a median unpaired one (the ocellar) arise from the upper third of the anterior face of the brain. The two optic nerves are the largest, arising very near the upper side of the brain, one on each side of the median furrow, so that the second and third sections made by the microtome, pass through them. Next below (from above downwards), is the origin of the single nerve sent to the two ocelli. We have not traced this nerve as far as the ocelli, but Milne-Edwards states that near the ocelli it divides into two branches. One of these two branches we figure in the drawing of the ocellus (plate 5. fig. 14). On each side of the ocellar nerve, and in nearly the same plane, arise two tegumental nerves, and directly below them a second pair of larger nerves (fronto-inferior tegumental) descend ventrally.

No nerves arise from the inferior half or two-thirds of the brain, which is smooth and rounded, with no indications of a median furrow.

It will thus be seen that, as stated by 'A. Milne-Edwards, there are no antennal nerves, such as usually exist in Arthropods with the exception of the Arachnida. This we have proved in the same manner as Milne-Edwards (though at the time ignorant that he had pursued the same method), by laying open with fine scissors the envelop (arterial or peri-neurial) which reaches to the posterior end of the brain, and seeing that the fibres of the nerves sent to the first pair of gnathopods originate quite independently of the brain itself. Moreover, after making sections of several brains, it is easy to see that only the commissures connecting the brain with the oesophageal ring are present; the nerves to the first pair of gnathopods not arising from the brain itself, but from the anterior and outer part of each side of the oesophageal ring, i. e., where the ring joins the brain; the commissure is very short in the larva, and obsolete in the adult.

Internal structure and histology of the brain. Most of the numerous stained and unstained transverse sections threw but little light on the topography; the nerve-fibres and ganglion-cells being apparently arranged horizontally, and mostly confined to the upper part of the brain; at any rate it was not until I had studied the horizontal sections, that I could gain an insight into the relation of parts as shown by sections cut vertically from in front backwards.

Finally a series of about fifty sections each, from two brains, cut by Mr. Mason horizontally from above, downwards, and carefully mounted in consecutive order, each section being numbered, has enabled me to arrive at a tolerably complete idea of the topography of the brain, so that I could mentally construct a model of the brain of Limulus, and compare it with the normal arthropod brain.

The histological elements are four in number: —

1. Large ganglion-cells, filled densely with granules, and with a well defined nucleus similarly filled and containing a granulated nucleolus. These cells (plate 7. fig. 3c) may be crowded or loosely grouped; the granular contents varying in density, and the walls of the cell thick and loose or thinner and dense; they terminate in large nerve fibres. They are similar in form and size, though not in topographical arrangement, to the large ganglionic cells of the lobster's brain (see plate 7, fig. 1b).

2. Smaller ganglion-cells, much more numerous than the larger, more hyaline, having much fewer granules and with the nuclei less distinctly outlined (plate 7, fig. 1c). They are seen to be somewhat smaller, but otherwise like those in the brain of the lobster, which we also figure (plate 7, fig. 1d).

3. Nerve fibres; these, like the large sized ganglion-cells from which they originate, are stained tawny yellowish brown with osmic acid. These fibres (plate 7, figs. 3a, 3b) are large and coarse, their fine granular contents homogeneous, and they closely resemble the nerve-fibres distributed to the compound and simple eyes of Limulus. Certain fibres near the origin of the optic nerves are distinctly nucleated at intervals (plate 7, fig. 3b).

4. Minute cells, or rather nuclei, very numerous and forming the large ruffle-like masses enveloped in connective tissue and constituting the greater part of the brain. They stain dark brown with osmic acid, so that these fungoid or ruffle-like bodies are readily distinguishable by their dark brown color from the surrounding tissues, which stain much lighter. In unstained sections simply hardened by alcohol, the tissue or bodies formed by these nuclei is darker than the other tissue, which is white. As these masses or bodies appear to be wholly made up of nuclei, I propose that they be distinguished by the name of *nucleogenous bodies*.

The brain itself is enveloped by a very thick, dense membrane, which I am disposed to regard as a neurilemma, homologous with that of the lobster's brain, though much thicker. It is formed of a fibrous connective tissue, and probably some elastic tissue, which directly penetrates into the brain-substance, forming a network of connective tissue enclosing the nucleogenous bodies; with occasionally clear nucleated portions in the spaces between the balls of minute nuclei, *i.e.*, the nucleogenous bodies. The fact that this envelop of the brain, a direct continuation of the so-called arterial coat of the oesophageal ring, is so intimately connected with the brain-substance itself, and that there is no space between it and the brain for the passage of the blood, and that an artery is situated outside of the brain (plate 7, fig. 4, *ar*), indicates strongly that this corresponds to the neurilemma of other invertebrates, or what Krieger designates as the " perineurium." It forms a fold on the upper side of the oesophageal ring, and thus becomes the direct continuation of the large lateral aortic branches; but it seems to be formed of short, tortuous fibres of connective tissue, with no true muscular fibres, such as are seen in transverse sections of the smaller arteries.

We will now describe the topography of the brain as seen in sections, beginning with the upper surface, at the origin of the optic nerves, and góing downwards. After the microtome has made five slices $\frac{1}{1000} - \frac{1}{500}$ inch thick, removing the upper part of the low elevations on each side of the broad, shallow median furrow, a section (plate 7, fig. 1) is obtained, which extends through the optic nerves, and also includes a part of the commissures by which the brain is connected with the oesophageal ring, the commissures being situated on the dorsal side of the central nervous system ; while the brain is here rather short antero-posteriorly, the median cleft or anterior end of the oesophageal opening projecting well into the brain ; the latter is more symmetrical in life than indicated in the figure, the brain being probably contracted a little in alcohol, and in the gum, while the razor made a slanting cut, so that while it passed through the middle of the right optic nerve, it merely grazed the edge of the opposite nerve.

At the uppermost region of the brain the sides are occupied by the nucleogenous bodies (*nb*), extending nearly to the back of the brain, but not reaching near the front. Within, but next to these nucleogenous bodies, are large sub-spherical masses of nucleated cells (fig. 1, *cm*), from which the optic nerves apparently arise. (Only the outer edge of the left mass has been cut through.) These cells are abundant and represented by fig. 2. They are hyaline, contain few granules, as do the nuclei. They are a little smaller than those of the lobster, which we have drawn to the same scale with the camera lucida, but it will be seen that they are identical morphologically. Enclosing in part the left mass of cells is a Y-shaped mass of fibres and nuclei (*y*) which reminds one of the trabeculae of the cockroach's brain. This apparently is not of much importance, and is an off-shoot from the central mass of nerve-fibres, as in five sections below, it merges with the other fibres. Just behind the middle of the brain, on each side of the median line, is a group of large ganglion-cells adjacent to the rounded cellular masses. Behind each group of large ganglion-cells originate the fibres of the commissures connecting the brain with the oesophageal ring; on the outside of the commissural nerve-fibres is a group of large ganglion cells longitudinally disposed. As we descend from the top to the base or under side of the brain, the commissure is cut through and disappears, the brain extending considerably below the oesophageal ring.

In the next section the large ganglion-cells are seen to be scattered through the middle of the posterior fibrous portion of the brain.

In the tenth section two large ruffle or fungus-like nucleogenous bodies appear, one on each side of the median line, with several smaller ones; and at the back part of the brain a bridge or transverse bundle of fibres now appears, connecting the mass of nerve-fibres on each side of the brain. This bridge becomes thicker as seen in the fourteenth section (plate 7, fig. 2).

This latter section passes through the lower edge of the right optic nerve (*op n*). The fibrous region is now invaded more than above by the nucleogenous bodies (*nb*), the former being mostly restricted to the posterior half of the brain, the brain itself being longer, and the oesophageal opening not extending so far into the middle of the brain. Its bilateral symmetry is seen to be tolerably marked. The Y-shaped fibrous mass is broad and obscurely marked; while the nucleogenous bodies occupy nearly two-thirds of the area of the section, the area having extended from the sides around to the front, nearly meeting on the median line of the brain. There are two central areas containing large ganglionic cells, and two other similar areas farther back and nearer the sides of the brain.

Plate 7, fig. 1*a*, represents the size of the large ganglion-cells of Limulus, compared with fig. 1*b* of a similar cell from the lobster's brain; the two being identical in size and in density of the protoplasmic granules.

Plate 7, fig. 3, represents a section through the front part of the right side of the brain; it shows the origin of the optic nerve from the small sized ganglion-cells in the central region of the brain. That the nerve-fibres within the brain are sometimes nucleated is shown by the adjoining figures (3*a*, 3*b*), where the nerves are cut transversely.

Plate 7, fig. 4 represents a section through the ocellar nerve, just grazing the right tegumental nerve. The fibrous portions are still more restricted, and they extend to the insertion of the nerves; the fibres are arranged into transverse as well as longitudinal bundles, of which the more important ones are figured, but of their origin and termination nothing definite has been ascertained.

Plate 7, fig. 5 represents a section through the middle of the ocellar and two tegumental nerves, and though it is obvious that the razor cut tolerably even, as the nerves are quite evenly severed, yet it will be seen how unsymmetrical the brain at this point is, after allowing for unequal contractions due to reagents. The median line between the two sides is very obscure and irregular; the mass of large ganglion-cells is quite large, and disposed in an unbroken mass on each side of the median line, should one be drawn through the brain. On the right side the fibres are almost wholly confined to an area near the middle of the right half, while the left side of the brain is mostly occupied with fibres, the nucleogenous bodies not extending to the back of the brain, as on the opposite side. At this part of the brain, in the more symmetrical specimen of the two brains specially studied, the nucleogenous bodies occupy nearly two-thirds of this plane of the brain; while the posterior group of large ganglion-cells is more extensive than above, and there are now but faint traces of the "bridge" of new fibres.

In a section lower down, near the middle plane of the brain, the nucleogenous bodies extend to the back of the brain, thus enclosing the mass of large ganglion-cells, which lies in front towards the middle of the brain; the nucleogenous bodies, at least the longer narrower masses, extend in towards the centre of the brain, so that they seem to radiate outwards from near the centre to the periphery.

In a section through the pair of lower tegumental nerves (fig. 6), in the same brain as represented at fig. 5, and on the same side, the fibrous masses are seen to be greatly reduced in extent, now filling up narrow spaces between the nucleogenous masses which converge towards the interior of the brain. The fibres evidently originate from the smaller and larger ganglion-cells, and pass forward and outward among the ruffle-like nucleogenous bodies. In the section here figured, the large ganglion-cells extend to the extreme back of the left side of the brain.

In another section below the nerves (fig. 7), the fibrous portion does not apparently reach the front, nor much beyond the middle of the brain, which at this point in one brain shows but slight symmetry, no fibres being visible in the right side.

Just below the section last figured, where no nerves are sent out from the brain, and before the sections diminish in size, the whole area seems to be filled with large rounded nucleogenous fungoid bodies, forming about eight irregular series passing from the back to the front of the brain, and arranged four on a side. A very few small bundles of nerve-fibres are to be seen, but with no determinate direction. This disposition of the histological elements extends downward to the bottom of the brain.

A transverse section of the brain from above downwards, cut just before the middle of the brain (fig. 8), shows nearly the same arrangement of parts as in horizontal sections; the upper part is seen to be occupied with the two larger groups of large ganglion-cells $(l\,g\,c)$, the nucleogenous bodies taking up most of the remainder of the brain, while a long bundle of nerve fibres (fa) passes from above downwards between the nucleogenous bodies.

To recapitulate and generalize from the foregoing facts: The brain is largely composed of masses of nuclei (nucleogenous bodies), enclosed by a mesh-work of connective tissue; these bodies nearly fill up the lower part of the brain, i. e., that part below the origin of the nerves. In the upper half or third of the brain whence the nerves originate, the larger and smaller ganglion-cells and bundles of nerve-fibres appear and preserve a more or less definite topographical relation to the entire brain. The nucleogenous bodies at and near the top of the brain are confined to each side of the brain, though masses of large ganglion-cells, associated with smaller ones, and nuclei, one on each side, just behind the middle, pass from below upwards; these groups of cells are more or less spherical as they grow smaller near the under side and at the top of the brain. The ganglion-cells altogether give rise to bundles of nerve-fibres; though it is probable that many nerve-fibres are without beginnings from cells, but originally developed from nuclei, as the ganglion-cells probably are in the beginning; since, in the larval brain, no fibres are to be seen, the brain substance consisting of cells alone. (See plate 3, fig. 3a.)

Thus the tract of nerve-fibres in each half of the brain is irregularly wedge-shaped, the apex situated near the centre of each hemisphere, and the base spreading out irregularly on the top, thus pushing aside, as it were, and crowding to the walls on each side the seemingly less dynamic portion of the brain, i. e., the masses of nuclei, or undeveloped cells (nucleogenous bodies). At the upper part of the back of the brain, just outside, at the origin of the posterior commissures, are two longitudinal groups of ganglion-cells on each side; these disappear below with the commissural nerves themselves.

The asymmetry of the brain, compared with that of other arthropods is remarkable; the large ganglion-cells are most abundant in the centre behind the middle, extending from that point to the posterior side of the brain; a median line is only slightly indicated by the arrangement of the fungoid bodies. The tract composed of large nerve-fibres, with scattered ganglion-cells on the left side, is much more extensive than on the right.

Comparison with the brain of other Arthropods. So wholly unlike in its form, the want of antennal nerves, and its internal structure, is the supra-oesophageal ganglion or brain of Limulus, to that of the higher Crustacea (i. e., Decapoda, the brain of the lower Crustacea not yet having been examined), that it is difficult to find any points of comparison.

Histologically, judging by my few sections of the lobster's brain which are stained with carmine, the brain of Limulus agrees with that of other arthropods in having similar large and small ganglion-cells, but the topography of the cell-masses essentially differs in the two types of brain. There are in Limulus no *Ballensubstanz*-masses, so characteristic of other arthropods, — the histological elements constituting these not having yet been discovered in Limulus.

We conclude, therefore, that, topographically, the internal structure of the brain of Limulus is constructed on a wholly different plan from that of any other arthropodous type known, so much so that it seems useless to attempt at present to homologize the different regions in the two types of brain. The plan is simple in Limulus; much more complex in other arthropods, especially in the brain of the decapodous, and probably most other Crustacea, the Decapoda having two pairs of antennal nerves beside the optic. In external appearance the two types of brain are entirely unlike. The symmetry of the brain of the crayfish and lobster and insects is beautifully marked (each hemisphere exactly repeating in its internal topography the structure of the opposite side), while that of Limulus is obscure and imperfect.

Structure of the oesophageal ganglia. (Plate 6, fig. 11.) A section through one side of the oesophageal ring, running through a ganglionic centre and the origin of the nerve to one of the anterior (second?) gnathopods (*gnn*) shows that the topography is quite simple. The central mass is mostly composed of nuclei and nerve fibres, the latter predominating until the nuclei disappear towards the base of the ganglion, where the nerve to the foot-jaw originates. On the outside of the ganglion, along nearly the whole length, are scattered large ganglion-cells (*lgc*). Near the upper and outer side is a group of small, narrow nucleogenous bodies (*nb*). There is a wide space for the passage of the blood between the ganglion with its nerve and the connective-tissue envelope, which is thick and of the same structure as the perineurium of the brain itself. This space extends along the whole length of the nervous system to the termination of the cord, the nerves sent to the appendages being enveloped by a continuation of the same coat. Among the large ganglion cells are numerous smaller ones, some of which are truly bipolar, as represented in our drawing (fig. 11*a*); the nuclei have distinct edges, so that I regard them simply as small-sized ganglion-cells rather than nucleated nerve-fibres.

Structure of the abdominal ganglia. There are six abdominal ganglia, the last being larger and longer than the others. A section through the second abdominal ganglion (plate 6, fig. 12) shows that the central mass of the double ganglion consists of longitudinal fibres, with scattered nuclei. On the upper side in the median line is a group of large and small ganglion-cells, and beneath is a mass extending to each side where they become most numerous. In some sections the central fibrous mass is enveloped by an irregular layer of ganglion-cells, some bipolar, with nerve-fibres forming a loose net work. In fig. 12 *a* a nerve connected by its neurilemma with that of the ganglion has been cut through; in this nerve there are only fibres present. In fig. 13 a large nerve leading to the abdominal appendage is seen to be sent off from one side of the double central mass; the other side (*gang*) has been torn away from the one opposite.

In neither this nor in sections of the last elongated abdominal ganglion were any nucleogenous bodies to be seen, so it seems most probable that none occur in the abdominal ganglia.

The section here figured of the last abdominal ganglion (plate 6, fig. 14) is seen to pass through four nerves, two on each side. The ganglion is seen to be formed by the union of the two separate cords, which are separate just before the ganglion. Above the ganglion on each side of the median line is a mass of large ganglion cells, of the same size as those of the brain, associated with more numerous smaller ones. This mass extends around and beneath each hemisphere of the ganglion, forming a layer of cells and fibres, some of the cells distinctly bipolar, which becomes interrupted at the median line, indicated by the deep notch in the central fibrous nerve-mass. The fibres from the laterally-situated cells are distinctly seen passing in and mingling with the fibres of the central nerve-mass; thus the nerves are reinforced from the peripheral ganglion cells. The central masses are composed of nerves, with a few nuclei; the fibres are mostly cut across, but occasionally short bundles of nerve fibres are seen lying across the cut ends of the others, though near the outer edge fibres are seen originating from the cells and passing in to the nerve mass.

FURTHER CONTRIBUTIONS TO THE EMBRYOLOGY OF LIMULUS.

The blastodermic skin or serous membrane. In my paper in the Memoirs of the Boston Society of Natural History I stated that the blastodermic skin, just before being moulted, consisted of nucleated cells, and I also traced its homology with the so-called serous membrane or outer "Faltenhulle" of the ectoderm of insects. In 1873,[1] by making transverse sections of the egg, I was able to study in a still more satisfactory manner these blastodermic cells, and to observe their nuclei before they became effaced during and after the moulting of the blastoderm.

On June 17th (the egg having been laid May 27th), the peripheral blastodermic cells began to harden, and the outer layer, that destined to form the outer or "serous" layer, to peel off from the primitive band beneath. The moult is accomplished by the flattened cells of the blastodermic skin hardening, and peeling off from those beneath. During this process the cells in this outer layer lose their nuclei, contracting and hardening during the process. Plate 3, fig. 14*a* shows at *o* the moulted empty cells with the nuclei empty and beginning to disappear, the walls being ragged and contracted; at *b* is the layer underneath of lining cells, with granules and distinct nuclei. Figs. 14*c* and 14*d* show the same cells during the moult, as seen from above and sideways; 14*b* represents the normal blastodermic cells, with a large, well-filled nucleus.

This blastodermic moult is comparable with that of Apus, as I have already observed,[2] the cells of the blastodermic skin in that animal being nucleated. This blastodermic skin may also, in its mode of development, be compared with the serous membrane of the scorpion as described by Metschnikoff, and with that of insects, in which at first the blastodermic cells are nucleated, and appear like those of Limulus. A similar moult takes place in Apus.

On June 19th, in other eggs, the cells of this membrane were observed to be empty, and the nuclei had lost their fine granules, and were beginning to disappear. The walls of the cells had become ragged through contraction, and in vertical sections short, peripheral, vertical, radiating lines could be perceived. At this time an interesting phenomenon was observed. In certain portions of the serous membrane the cells had become effaced, transitions from the rudiments of cells to those fully formed being seen. In insects and crustaceans, as a rule, the cells all finally disappear, the serous membrane being structureless and homogeneous. The relation of the blastodermic cells in the serous membrane of Limulus is due, without doubt, to the singular function this skin is destined to perform; i. e., its use as a vicarious chorion, the chorion itself splitting apart and falling off in consequence of the increase in size of the embryo.

---

[1] The substance of this account appeared in the American   [2] See Memoirs Bost. Soc. Nat. Hist., II, 161, foot-note.
Naturalist, Nov. 1873, VII, p. 675.

Development of the internal organs. Although a good many eggs were sliced, I was unable to discover any in the stage when the ectoderm and endoderm are differentiated, nor to examine the embryo in the gastrula condition, if there be such. The eggs were either in the stage of segmentation of the yolk, or the embryo was so far advanced that the indications of the segments had appeared. This period of development of the gastrula is evidently intermediate between the stages, plate 3, fig. 7, and fig. 10 of my first memoir The succession in which the more important system of organs arise, is as follows : — first the nervous system ; long afterwards the muscles and the heart. These organs are well developed before the larva hatches, though the first indications of the mesoderm were not observed. It is not for some time after hatching that the digestive canal as a whole is formed ; although in the gastrula condition an archenteron may probably be developed, I have been unable to detect, after making numerous sections of eggs and embryos, any traces of the stomach and intestine until long after the larva has hatched. The primitive liver-tubules and the ovaries seem to arise at about the same time after the digestive canal is indicated. The development of the renal organs was not traced, no indications of these organs being detected.

The eyes begin to form at the time of hatching, before the digestive tract is indicated. But little attention was devoted to the mode of development of the compound eyes. They are then very small black spots, the rudimentary corneal lenses few in number, and conical. The black retina is underlaid by a white mass ; plate 4, fig. 4 represents one of the ocelli at or soon after exclusion from the egg ; the external region is clear and made up of about twenty elongated epithelial cells, with a distinct refractive nucleus and granules ; whether these are pigment cells or not we did not farther observe ; underneath this area is the dark pigment mass in which no cells could be detected with a $\frac{1}{5}$ objective and B eyepiece ; the ends of the epithelial cells seem to sink into the mass.

Development of the nervous system.[1] After a number of unsuccessful attempts at discovering the first indications of the nervous system, I at length discovered, in thin sections kindly made for me by Prof. T. D. Biscoe, the nervous tract in a transverse section of an embryo in an early stage of development, corresponding to that figured on plate 6, fig. 10, of my first memoir. The period at which it was first observable was posterior to the first blastodermic moult, and before the appearance of the rudiments of the six pairs of cephalothoracic limbs (gnathopods). The primitive band now entirely surrounds the yolk, being much thicker on one side of the egg than on the other, the limbs budding out from this disk-like, thickened portion, most of which represents the ectoderm. At the time the nervous cord was observed it was entirely differentiated and quite distinct from the surrounding tissue of the ectoderm.[2]

At a later stage in the embryo, represented by plate 5, fig. 16, in my first memoir, at a period when the body is divided into a cephalothorax and abdomen, and the limbs are developed, by a series of sections made parallel with the under surface of the body, I could

---

[1] The principal points in this section were originally printed in a short notice in the American Naturalist, July 1875, IX, 422.

[2] Plate 4, fig. 3, represents the nerve cells, and fig. 3a, the cells of the mass of connective tissue in which the two cords are embedded, from a freshly hatched larva.

make out the general form of the main nervous cord.   Plate 3, fig. 3, shows the general
relations of the cord to the body.   It is large and broad, with three well-marked pairs
of consolidated ganglia in the abdomen, the two basal ones supplying the nerves for the
first and second abdominal feet.   There are in the cephalothorax six pairs of consolidated
ganglia, the commissures being as yet undeveloped; the ganglia are indicated by the
minute openings in front of and behind each pair of ganglia.   The ganglia of the first pair
of feet could be clearly distinguished; the brain or cephalic ganglion is probably represented at fig. 3 I; fig. 3 α, the same enlarged.   The number of ganglia, throwing out the
brain, is nine, corresponding to the six pairs of cephalothoracic feet and the two abdominal
segments, there being at this stage but two pairs of appendages in the abdomen.

The next important stage of development is seen in longitudinal sections of the larva
after hatching, and when the digestive canal is marked out.   To show the ganglia best,
the section should be made on one side of the median line of the body, so as to pass
through the middle of the ganglia on one side.   Plate 3, fig. 2, shows a section thus made
and stained with carmine ; the nervous ganglia remaining white are very clearly indicated ;
the commissures are not shown, but they are now developed, since the ganglia are mostly
separate.

Now if we make a longitudinal section of the young horse-shoe crab when a little over
an inch long, the disposition of the nervous cord is exactly as in the full grown individual,
as figured by A. Milne-Edwards; see also our representation on plate 3, fig. 1, br, oe r.
The nervous ganglia are then united into a nearly continuous nervous collar, the opening
in front being filled up by the brain or cephalic ganglion.[1]

Turning now to the nervous system of the larva (plate 3, fig. 2), the section here figured
shows a most important and interesting difference as regards the ganglia which supply
nerves to the appendages of the cephalothorax.   They are at this time entirely separate,
the spaces between the four posterior ones, which are connected by commissures, being as
wide as the ganglia themselves are thick.   There are behind the oesophagus six ganglia,
corresponding to each of the six pairs of gnathopods; while the brain is rather larger
than the others, and the first post-oesophageal ganglia are the smallest of the six, corresponding to the more diminutive size of the first pair of gnathopods.

Reference may also be made here to plate 5, fig. 8, which shows the mode of origin of
the nerves distributed from the first post-oesophageal ganglion to the feet; this section
certainly very clearly demonstrates that the first pair of gnathopods belong with the postoral series, that they can in no sense be regarded as homologues of the antennae of
other Arthropods, and that in fact there are no antennae in Limulus, and without doubt
in the Merostomata in general.   But this subject has been already discussed in the
chapter on morphology.

It is not until after the second moult that the adult condition of the nervous system is
attained, as Dr. Dohrn[2] has figured the separate ganglia in a larva which had evidently
moulted once, the abdominal spine being well developed.   This is certainly an interesting

[1] For the nature of the brain and the oesophageal ganglionic collar, the reader is referred to the section of this
paper on the structure of the adult brain.

[2] Dohrn. Zur Embryologie und Morphologie des Limulus
polyphemus. He also represents the fourth pair of abdominal
appendages ; the larva has but three before the first moult.

instance of the metamorphosis and cephalization of the nervous system, which is carried
on internally, though the other organs and outer body-form remain unchanged.

Development of the digestive canal. Unfortunately the mode of formation of the primitive digestive cavity or archenteron was not observed, as eggs showing the formation of
a gastrula could not be obtained. From this early period until after the larva has
hatched the entire canal remains unorganized, the entire body-cavity being filled between
the heart and nervous tract with yolk granules.

The earliest stage when the enteric canal was observed at all was after the different
parts — oesophagus, crop, stomach, intestines, and cloaca or rectum — had assumed their
definitive shape. Plate 4, fig. 2, illustrates a section of the larva before its first moult,
through the head. The space around the heart and digestive canal and over the nervous
cord is filled with a very loose connective tissue; the cells, which are nucleated, spindle-
shaped or triangular, being scattered, and forming a very open net-work of cells. In after-
life the cells multiply, becoming very numerous and round or oval in form. This connective tissue extends throughout the entire body-cavity, the ovarian or testicular tubes
ramifying throughout the mass, as well as the liver tubules.

The section at plate 5, fig. 8, passes through the oesophagus and the crop. The former
(figs. 10, 11, enlarged) is apparently filled with a few large epithelial cells, which represent
the folds of the lining of the oesophagus. The walls of the proventriculus are very thick;
the lumen or passage is lined with the alternating larger and smaller folds of spherical
epithelial cells, and with a thin semi-chitinous layer; the muscular layer, representing the
endoderm built up around the originally invaginated ectodermal layer forming the fore gut
or protenteron (plate 5, figs. 7, 7*a*), shows the epithelium of the intestine, the cells being
very irregular in size and length.

Origin of the liver. Plate 4, fig. 7, represents a section through the middle of the
cephalothorax, passing through the intestine and one of the pairs of biliary ducts. The
ducts are seen to open directly into the stomach, the duct being large, and at first there
is a primary liver-tube, which bends downward at quite an angle before passing to the
outer edge of the carapace. There are thus four primary biliary tubes, these in after
life subdividing and ramifying throughout the body-cavity to an indefinite extent. The
tubes are clear, transparent, with dark granules.

Development of the ovary. The same section represented in plate 4, fig. 7, also passes
two bodies, one on the outer side of and just below the heart, on each side of the mid-gut.
These are the rudimentary ovaries. One section (fig. 8) shows the ovarian follicles
attached to the walls of the gland, and, in fig. 8*a*, the ovarian eggs are just beginning to
form, constituting a mass apparently free from the walls of the ovarian tubules.

Structure of the testes and development of the spermatozoa. In our first memoir on Limulus we figured the spermatozoa; since then Professor Lankester has also described them.

The argument that Limulus is not a Crustacean because the spermatozoa have tails is
somewhat vitiated by the fact that those of the barnacles have exceedingly long, well

developed tails. We introduce a figure of those of a species of Lepas collected at Penikese, an island at the mouth of Buzzard's Bay. The head is broad and flat, plainly sinuous seen sidewise, the centre being filled with granules. The spermatocysts (plate 3, fig. 5) are spherical, usually containing five spermatozoa.

The tubules of the testis of Limulus are yellowish ; this color is due to the presence of numerous yellow pigment granules. Fig. 4a represents the epithelial tissue forming the walls of the follicles (fig. 4). The spermatocysts (fig. 4d) are spherical, containing four immature spermatozoa. while the earlier condition of the same is seen at fig. 4b, where the sperm-cells are nucleated.

We introduce for comparison drawings (fig. 6) of the spermatocysts of a decapod Crustacean (*Libinia canaliculata*), the spermatozoa (fig. 6f) being tailless and nucleated. Certain larger cells have a large nucleus, with a small nucleolus ; the nature of these cells we do not understand.

BIBLIOGRAPHY.

VAN DER HOEVEN. Recherches sur l'Histoire Naturelle et l'Anatomie des Limules, par J. Van der Hoeven. Avec sept planches. Leyde, 1838. fol. pp. 38.

GEGENBAUR. Anatomische Untersuchung eines Limulus, mit besonderer Berücksichtigung der Gewebe. Von C. Gegenbaur. Mit einer tafel. Besonders abgedruckt aus dem 4. Bande der Abhandlungen der naturforschenden Gesellschaft in Halle. Halle, 1858. 4°. pp. 24.

LOCKWOOD. The Horse-foot Crab. By Rev. S. Lockwood. American Naturalist, IV, July, 1870.

PACKARD. The Embryology of Limulus polyphemus. By A. S. Packard, Jr. American Naturalist, IV, pp. 257-274. July, 1870. American Naturalist, IV, pp. 498-502. October, 1870. Proceedings Bost. Soc. Nat. Hist., June, 1871. Vol. XIV, p. 60.

————— On the Embryology of Limulus polyphemus. By A. S. Packard, Jr. Proceedings American Association Adv. Science, 19th Meeting, Troy, N. Y., July, 1871. 8°. pp. 8.

————— Morphology and Ancestry of the King Crab. By A. S. Packard, Jr. American Naturalist, IV, pp. 754-756. Feb., 1871.

————— The Development of Limulus polyphemus. By A. S. Packard, Jr. Memoirs Bost. Soc. Nat. Hist., II, pp. 155-202. March, 1872.

————— Further Observations on the Embryology of Limulus, with Notes on its Affinities. By A. S. Packard. Amer. Nat., VII, pp. 675-678. Nov., 1873. Proceedings Amer. Assoc. Adv. Science. Portland Meeting, 1874.

————— On the Development of the Nervous System in Limulus. By A. S. Packard, Jr. Amer. Nat., IX, pp. 422-424. July, 1875.

————— On an Undescribed Organ in Limulus, supposed to be renal in its nature. By A. S. Packard, Jr. Amer. Nat., IX, pp. 511-514. Sept., 1875.

————— Structure of the Eye of Limulus. Amer. Nat., XIV, pp. 212-213. March, 1880.

————— Internal Structure of the Brain of Limulus. Amer. Nat., XIV, pp. 445-448. June, 1880.

VAN BENEDEN. De la place qui les Limules doivent occuper dans la Classification des Arthropodes d'apres leur developpement embryonnaire ; par Édouard van Beneden. Communiqué à la Soc. Ent. de Belgique, 14 Oct. 1871. Gervais, Journ. Zoologie, I, 1872, pp. 41-44. Annals and Mag. Nat. Hist., 1872.

DOHRN. Untersuchungen über den Ban und Entwickelung der Arthropoden. Von Anton Dohrn. Abdruck aus der Jenaischen Zeitschrift Wissensch., Band VI, Heft 4. pp. 582-640. 1871.

MILNE-EDWARDS. Recherches sur l'Anatomie des Limules. Par A. Milne-Edwards. Annales des Sciences Nat., XVII, pp. 67. 11 plates. Nov. 1872. Commission Scientifique du Mexique.

OWEN. Anatomy of the King Crab. (Limulus polyphemus Latr.) By R. Owen. London, 1873. Trans. Linn. Society, London. 5 plates. 4°. pp. 50.

LANKESTER. Mobility of the Spermatozoids of Limulus. By E. R. Lankester. Quart. Journ. Micr. Science. Oct., 1878. pp. 453–454.

See also Strauss-Dürckheim's Traité d'Anatomie Comparative. 1842.

Owen's Lectures on the Invertebrate Animals. 1843–1855.

Woodward's papers on Merostomata. Palaeontol. Society. 1866–1878.

Huxley's Anatomy of the Invertebrate Animals. 1877.

W. Grenacher's Untersuchungen über das Sehorgan der Arthropoden. 1879.

EXPLANATION OF THE PLATES.

PLATE I.

Fig. 1. Under side of a Limulus, a little over two inches long without the spine, injected to show the abundance of the arterial twigs in the limbs and caudal spine as well as the body. The injection was made at Penikese by the late Edwin Bicknell. J. S. Kingsley, del.

Fig. 2. Camera lucida drawing of a living larva, showing the circulation of the blood-corpuscles in the right under-side of the abdomen and on the left first abdominal limb. Author, del.

Fig. 3. Camera lucida drawings, showing the actual course taken by the blood corpuscles in the first abdominal appendage of the same larval Limulus. The arrows show the direction of the currents of blood, with the corpuscles; the blood passing from the heart down along the inner side of the appendage, and passing by tortuous, irregular courses around by the outside, back along the base, and returning to the pericardial chamber through the venous opening. This mode of circulation is much as we have seen take place in the amphipodous Crustacea. Author, del.

PLATE II.

Fig. 1. Transverse section of adult male Limulus, natural size, through the proventriculus [*pr*], showing the cone [*c*], the oesophagus [*oe*], and the brain [*br*]; *a*, aorta, or frontal artery; *col*, collective venous sinus. From a drawing made for the author at Penikese by P. Roetter.

Fig. 2. Section through the cephalothorax in front of the heart, brain, and first pair of gnathopods; *m*, muscles. J. S. Kingsley, del.

Fig. 3. Section through the cephalothorax behind the first pair of gnathopods; *ht*, heart; *m*, great longitudinal adductor muscle; *cp*, supraneural cartilaginous plate protecting the central nervous system. The latter not shown. J. S. Kingsley, del.

Fig. 4. Transverse section through the abdomen, showing the second abdominal, or first respiratory, foot; *ht*, heart, beneath which is the intestine; *b*, origin and middle of branchio-cardiac veins, which carry the blood from the limb to the heart. J. S. Kingsley, del.

PLATE III.

Fig. 1. Longitudinal section through a Limulus about two inches long, exclusive of the caudal spine; *ht*, heart; *m*, mouth, leading by the oesophagus to the proventriculus [*pr*]; *cone*, proventricular cone; *st*, stomach; *in*, intestine; *a*, anus; *br*, brain, or supraoesophageal ganglion, behind which is a part of the oesophageal ring [*oe. r*]; *ng*, ganglionated cord; *ct*, supraneural cartilaginous plate; enlarged about twice.

Fig. 2. Longitudinal section through the larva of Limulus on one side of the heart and digestive tract, passing through the brain and cephalothoracic ganglia; *br*, brain; the six other ganglia [1–6] separate from one another, and afterwards consolidate to form the "oesophageal ring"; 1, the first ganglion which supplies a pair of nerves to the first pair of gnathopods. [Compare plate 4, fig. 7, *gn*.]

Fig. 3. A horizontal section through the embryo long before it hatches, before the body has become flattened, before the heart and digestive canal have appeared, and soon after the embryo has reached the stage represented by plate 4. figs. 19, 19*a*, of our first memoir. There are six cephalothoracic ganglia [I–VI] besides the brain, and three abdominal ones [I–III]; the first two abdominal ones corresponding to the rudiments of the first and second abdominal appendages. 1–6, the six pairs of gnathopods; I, II, the two pairs of abdominal legs.

Fig. 3*a*. Enlarged view of the brain [?], the nerve cells [*b*] forming the ganglion, which is enveloped by connective tissue cells [*ct*]. (Is it these latter which are destined to form the nucleogenous bodies of the adult brain?)

Fig. 4. Follicles at end of a seminal tubule of testis of Limulus; 4 *a*, epithelial cells of seminal tubules, nucleated and highly refractive; × Tolles' $\frac{1}{5}$ objective C eye piece, magnified 725 diameters; 4*b*, amber-colored pigment cells of testis; 4*c*, similar but larger cells; 4*d*, spermatocysts of Limulus; 4*e*, cells associated with the spermatocysts, with a large nucleus and a distinct nucleolus; × Hartnack No. 9, B eyepiece.

Fig. 5. Spermatocyst of a barnacle [Lepas], 5*a*, side view, and 5*b*, front view, of a spermatozoon of the same; × $\frac{1}{10}$ B.

Fig. 6. Spermatocysts of different shapes, *a*, *b*, *c*, *d*, *e* (× $\frac{1}{5}$ B), and [*f*] tailless spermatozoon of *Libinia canaliculata*; × Hartnack No. 9.

Fig. 7. Supposed renal glands of Limulus; *b*, one of the four lobes extending upwards from the main stem [*a*]; *c*, chitinous bases of the gnathopods. 7*a*, reddish pigment bodies coloring the cellular mass of the gland, the cells being nucleated. 7*b*, 7*c*, two amber-colored yellow secreting cells scattered through the cellular mass, composed of nucleated cells, as at 7*a*; × Hartnack No. 9. B.

Fig. 8. Tubules of liver of living Limulus: × 30 diameters; 8*a*, a parent cell of the smaller liver cells; the shaded ones horn-colored, those unshaded clear; 8*b*, free liver cell; 8*c*, the same with pale nuclei. 8*d*, liver cells of Panopaeus; × $\frac{1}{5}$ Tolles B.

Fig. 9. Sections of liver tubes stained with carmine; × $\frac{1}{2}$ A.

Fig. 10. End of a liver tubule of *Homarus americanus*; × $\frac{1}{4}$ B.

Fig. 11, 11*a*. Striated muscle near insertion of leg of Limulus; × $\frac{1}{5}$ C (725 diameters).

Fig. 12, 12*a*, 12*b*. Sections through minute peripheral arteries near the compound eye; × $\frac{1}{2}$ A.

Fig. 13. White fibrous cartilage of the supraneural cartilaginous plate; longitudinal section showing the fibres on one edge and the nucleated cells in the dense structureless portion.

Fig. 14. Portion of the blastoderm lying next to the chorion [*ch*] with yolk granules; 14*a*, the same after the outer layer [*o*] has begun to moult, the cells beginning to wrinkle on the edges, and being without the protoplasmic granules [14*b*] seen in the deeper layer of blastodermic cells; 14*c*, vertical, and 14*d*, profile view of the same cells after moulting, the walls contracted and wrinkled, and with the nuclei partly absent or absorbed; × $\frac{1}{5}$ A.

Plate IV.

Fig. 1. Section through the larva some time after hatching; *ht*, heart; *int*, intestines; *nc*, double nervous cord; the muscular system well developed; *am*, undeveloped adductor muscle. The parenchym of the body consists of incipient connective tissue and liver-cells.

Fig. 2. Section through the cephalothorax of the same larva as represented at Fig. 1, the section passing through the compound eye [*ce*], the heart [*ht*], proventriculus [*pv*], and the double nervous cord; as yet the neurilemma is unformed, the nervous cord not being enveloped by it, this being represented by connective tissue [*ct*].

Fig. 3. Nerve cells of nervous cord of a freshly hatched larva, before the digestive tract and heart are indicated; 3*a*, connective tissue cells enveloping the nervous cord of 3; from these cells the neurilemma is probably formed.

Fig. 4. An ocellus of a larval Limulus, showing the epithelial cells [*e*] and the dark pigment of the retina [*r*]; × $\frac{1}{5}$ B. The ocelli are at this stage quite far apart.

Fig. 5. Section of nervous cord [*n*] embedded in connective tissue [*ct*], the section passing through the body near the eyes of an advanced larva, in which the heart and digestive tract are developed.

Fig. 6. Section through a ganglion [*g*] of the same larva as represented in Fig. 5, the ganglion completely surrounded by the connective tissue [*ct*].

Fig. 7. Section through an advanced larva showing the origin of a pair of liver ducts from the intestines [*int*], and a single primitive liver-duct [*ld*], of which there are two pairs; *ht*, heart; *gn*, a pair of nerves sent from the ganglion [*g*] to each second gnathopod.

Fig. 8. Section of an ovarian tube, with the ovarian follicles on the side; 8*a*, another section showing the cell-eggs.

PLATE V.

Fig. 1. Section through the vertical folds or teeth of the fore part of the crop or proventriculus; *m*, muscular layer; *pe*, pavement epithelium; *ce*, columnar epithelium; *ch*, chitinous layer.

Fig. 2. The central tooth of Fig. 1 magnified; × ½ A; lettering as before.

Fig. 3. Columnar epithelium from section of end of the oesophagus.

Fig. 4. Nucleated cells and fibres of the pavement epithelium of intestine; × ½ B; 4*a*, the same somewhat enlarged.

Fig. 5. Section from posterior part of the oesophagus, showing the chitinous layer [*ch*]; the empty spaces in the lobes surrounded by columnar epithelium [*ce*]; the pavement epithelium [*pe*] supporting the former.

Fig. 6. Pavement epithelium of rectal folds.

Fig. 7, 7*a*. Section of stomach of larva where the chitinous lining is absent, showing the irregularity of the epithelium.

Fig. 8. Section through an advanced larval Limulus, the figure indicating only the portion lying under the central lobe; *ht*, heart; *pr*, proventriculus; *oe*, oesophagus; *g*, first pair of ganglia, the oesophageal ring not yet being consolidated; *gn*, nerve to the first gnathopod [*gp*], demonstrating that the brain does not supply the nerves to the first pair of feet; *ct*, connective tissue, the neurilemma not yet formed.

Fig. 9. Section of inner part of the proventriculus showing the larger teeth [*t*] alternating with the smaller ones [*t*]; × ½ A.

Fig. 10. Section through the oesophagus; × ½ A.

Fig. 11. Another section of the same.

Fig. 12. Section through the simple eye or ocellus of Limulus; 3, third layer of the integument, clear and laminated; × ½ A; 2, second layer of integument finely granulated and laminated; *pe*, pore canals filled with connective tissue [*ct*]; *cl*, corneal lens; *h*, cup-shaped depression in the base of the corneal lens.

Fig. 13. Another section of an ocellus more enlarged; lettering as in fig. 12.

Fig. 14. Section through an ocellus showing the relations of the ocellar nerve and its branches [*ocn*]; 1, first and outer clear layer of the convex cornea; 2, second layer, finely laminated; 3, third, clear layer, with a few laminæ; *rt*, pigment layer in retina; *h*, hypodermis, of which the retina is a modification; *cl*, corneal lens; *ct*, connective tissue.

PLATE VI.

Fig. 1. Section through the entire compound eye of Limulus, stained with picro-carmine, showing the relations of the cornea and corneal lenses and retina to the branches of the optic nerve; *cor*, cornea; 1, outer clear, 2, middle laminated, and 3, inner clear portion of the chitinous cornea, seen to extend into the integument; *pc*, pore or nutritive canals filled with connective tissue; *cl*, corneal lenses; *rt*, retina; *hy*, hypodermis, of which the retina is a modification. Below is the mass of connective tissue cells [*rt*], through which the tortuous branches of the optic nerve pass and impinge on the ends of the conical corneal lenses; owing to the tortuous course of the nerve-fibres, they appear not to be continuous in the thin section of which this is a drawing. *ov*, ovary with cell eggs; *ar*, two arterial twigs; *l*, two liver tubes; *ict*, inner, darker brown connective tissue of the interior of the cephalothorax.

Fig. 2. Sections of corneal lenses in the middle of the eye; the retina has been removed by acid; *cl*, corneal lense: *h*, cup-shaped depression in base of lens; 2*a*, the same from near the periphery of the eye, where the corneal lenses are longer and more oblique directed inwards towards the middle of the eye.

Fig. 3. Epithelium of the retina around the end of a cone; *rhab*, rhabdom; *rcl*, retinal cells.

Fig. 4. Section of two retinulas, with the rhabdom [*rhab*] in the centre; × ½ A; 4*a*, a retinula [*ret*] od with acid to show the twelve cells into each of which a ray of the rhabdom projects; × ½ B.

Fig. 5. Section of soft parts of ocellus of Limulus, showing the subdivisions and mode of termination of one branch of the ocellar nerve [on]; the branches are enveloped in connective tissue [ct]. The section passes on one side of the corneal lens.

Fig. 6. Optical section of corneal lenses of Limulus, as seen through the transparent cornea, showing their slightly hexagonal appearance; × 50 diameters.

Fig. 7. Artificial section through the eye of Asaphus, a trilobite, to show the close similarity to the corneal lenses of Limulus; × ½ B eye-piece.

Fig. 8. Longitudinal section through the eye of Asaphus showing the corneal lenses; × ½ A. (Compare with Fig. 1, 2, 2a, the corneal lenses of Limulus.)

Fig. 9. Longitudinal section through the eye of Asaphus gigas; cl, corneal lenses; pc, pore canal; rt? probable indications of the upper edge of the retina?

Fig. 10. Section of part of the cornea of an Asaphus gigas which has been broken, showing several entire corneal lenses side by side.

Fig. 11. Section through one side of oesophageal ring passing through the nerve to one of the gnathopods or cephalothoracic feet; nl, neurilemma; sp, space between neurilemma and the ganglion; nb, small nucleogenous bodies at top of section or upper side of oesophageal ring; lgc, longitudinal group of large ganglion cells, extending along the outside of the oesophageal ring; gn n, gnathopodal nerve ; magnified 30 diameters.

Fig. 11a. A large ganglion cell [lgc], surrounded by smaller bipolar ganglion cells, magnified 224 diameters.

Fig. 12. Section through second abdominal ganglion, n, nerve to one of the abdominal feet; fa, fibres of central nerves; lgc, layer of large ganglion cells and nerve fibres arising from them; × 30 diam.

Fig. 13. Section through the same ganglion showing origin of nerve [n] to second abdominal foot.

Fig. 14. Section through the sixth or last abdominal ganglion passing through the nerves [n]; nf, shows the nerve fibres arising from the large ganglion cells and reinforcing the nerves making up the central mass, which is seen to be composed of the union of two separate nervous ends; × 30 diameters.

Fig. 15. Transverse section through the middle of the brain, showing the arrangement of the fibres [fa], nucleogenous bodies [nb] and groups of large ganglion-cells [lgc].

PLATE VII.

Fig. 1. Section through upper part of brain of Limulus, passing through the optic nerves [op n]; c m, groups of cells from which the optic nerves appear to arise; y, Y-shaped bundle of nerve-fibres; n b, nucleogenous bodies on each side of the brain; l g c, groups of large ganglion-cells; cm, commissures uniting the brain with the oesophageal ring; n l, neurilemma. The lettering the same for all the figures. Magnified 15–20 diameters. 1a, A large ganglion cell of Limulus; 1b, the same of the lobster; 1c, small ganglion cells of Limulus; 1d, the same of the lobster; all magnified 225 diameters to show their relative size and form.

Fig. 2. Section lower down, just grazing the under side of one optic nerve; the nucleogenous ruffle-shaped bodies in front as well as on the sides; the Y-shaped bundle of nerves nearly merged with the rest of the fibrous portion; the groups of large ganglion-cells [lgc] limited in extent.

Fig. 3. Section of portion of brain magnified 30 diameters showing the origin of left optic nerve ; f, bundle of nerve-fibres, without cells and nuclei; n¹, a nerve of which an enlarged section is seen at fig. 3a, showing the nucleated fibres cut across; n², a large bundle of nerve-fibres, of part of which, fig. 3b, is an enlarged view, showing the nucleated fibres in section, and seen longitudinally with a few nuclei visible; × 225 diameters; 3c, a group of large ganglion-cells, with branched nerve-fibres arising from them; n, nucleus; nc, nucleolus, magnified 225 diameters; 3d, a single large ganglion cell, giving origin to a branching nerve.

Fig. 4. Section of right side of brain passing through the ocellar nerve, o c n; c, commissure with large ganglion-cells and fibres at this point, surrounded by a distinct neurilemma; ar, artery passing down the back of the brain. Magnified 80 diameters.

Fig. 5. Section of brain of Limulus through the ocellar nerve [o c n] and the two tegumental nerves [t n]; c, section of lower part of commissure to oesophageal ring; fa¹ the small area on the right side composed of nerve-fibres, showing the asymmetry of the brain; magnified 80 diameters.

Fig. 6.   Section of left side of brain of Limulus, below the ocellar nerve, passing through the lower set of tegumental nerves.   The fibrous area [ƒ a] much branched, and still much greater than on the right side [ƒ a'].   A number of large ganglion cells are present at the posterior outer portion of the brain.   Magnified 30 diameters.

Fig. 7.   Section through left side of brain of Limulus, below any of the nerves, quite near the base of the brain, and showing how much the nucleogenous bodies have encroached on the fibrous area [ƒ a].

Figs. 1 and 2, were cut from the same brain; figs. 3, 4, 5, 6 and 7 from another brain; and were selected from about ninety other sections.

NOTE.—All the figures in plates 1–7 were drawn by the author, except plate 2, fig. 1, drawn by P. Roetter ; and plate 1, fig. 1, and all the figures on plate 2 (except fig. 1), which were drawn by J. S.Kingsley.

ZOOLOGICAL LABORATORY OF BROWN UNIVERSITY,
Providence, R. I., May, 1880.

ERRATA.

Page 10, line 30, for Arthopoda *read* Arthropoda.
Page 16, line 1 from bottom of footnote, for *Tottennia read Tottenia.*
Page 23, line 25, for (Fig. 9, *rt*) *read* (Fig. 9, *rt* ?).
Page 26, line 6 from bottom for " or lens-epithelium " *read* "lens-epithelium."
Page 29, line 14, for analagous *read* analogous.
Page 29, line 37, after the word but *dele* that.

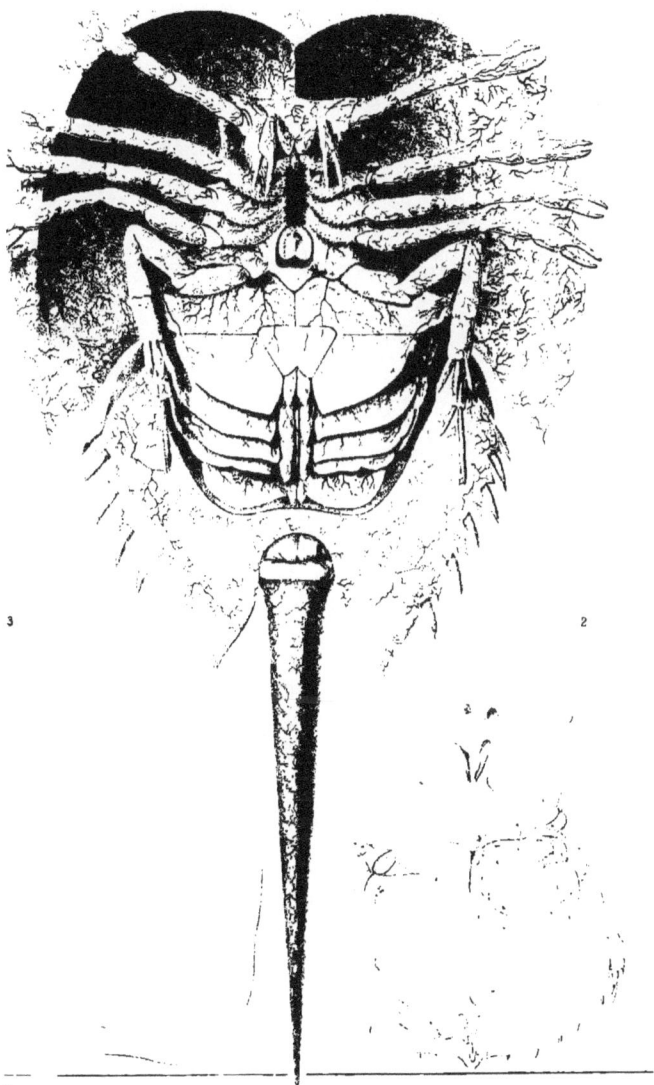

CIRCULATION OF THE BLOOD IN THE KING CRAB

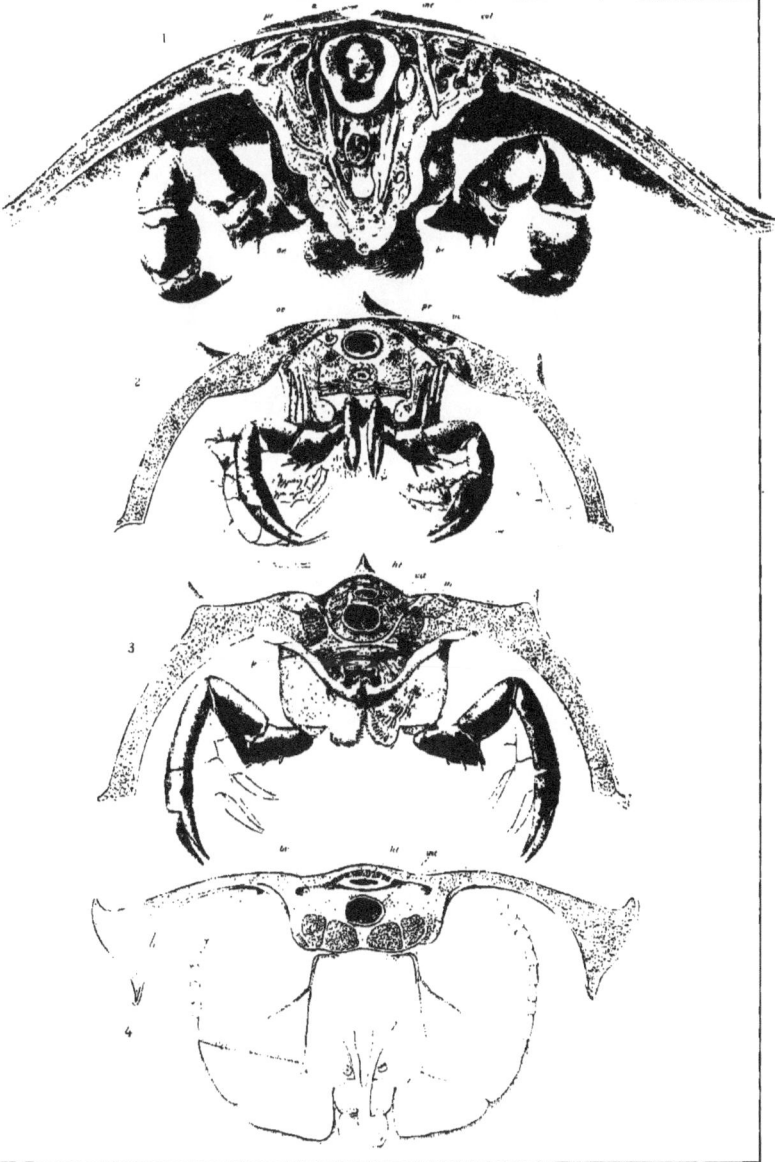

SECTIONS THROUGH ADULT KING CRAB.

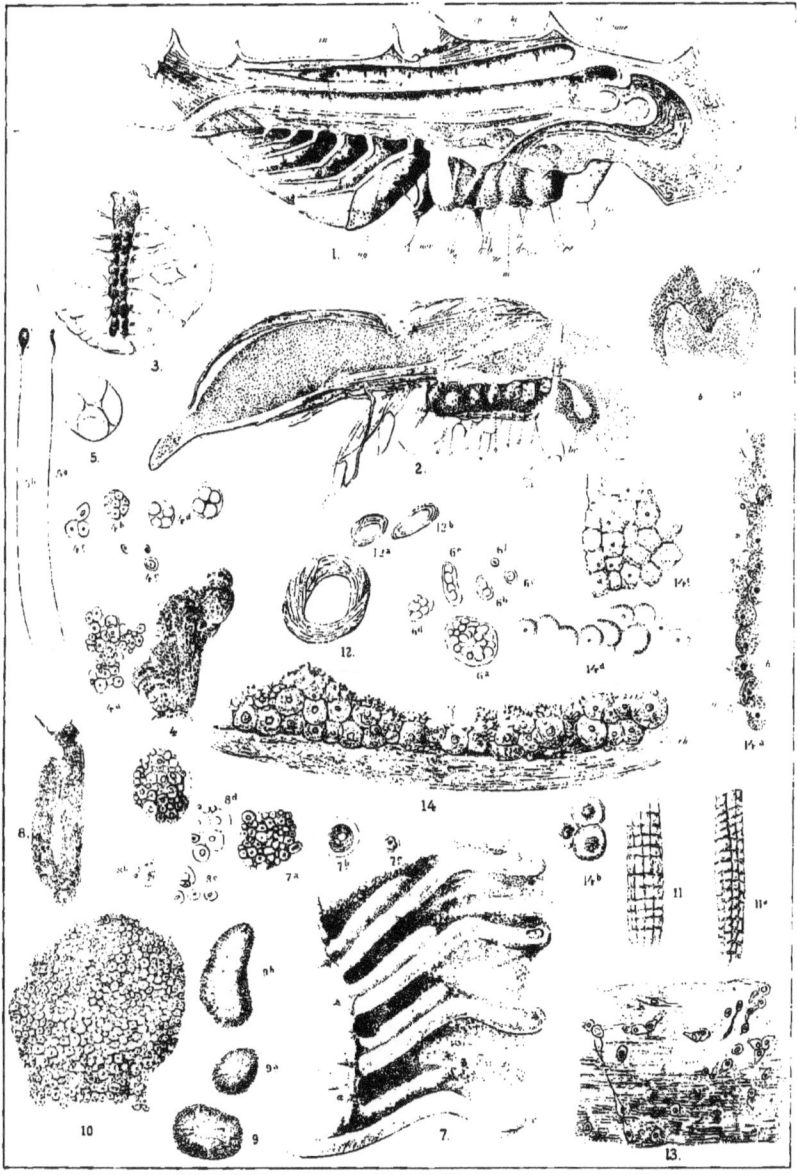

A S Packard del

A Meisel lith

ANATOMY AND HISTOLOGY OF THE KING CRAB

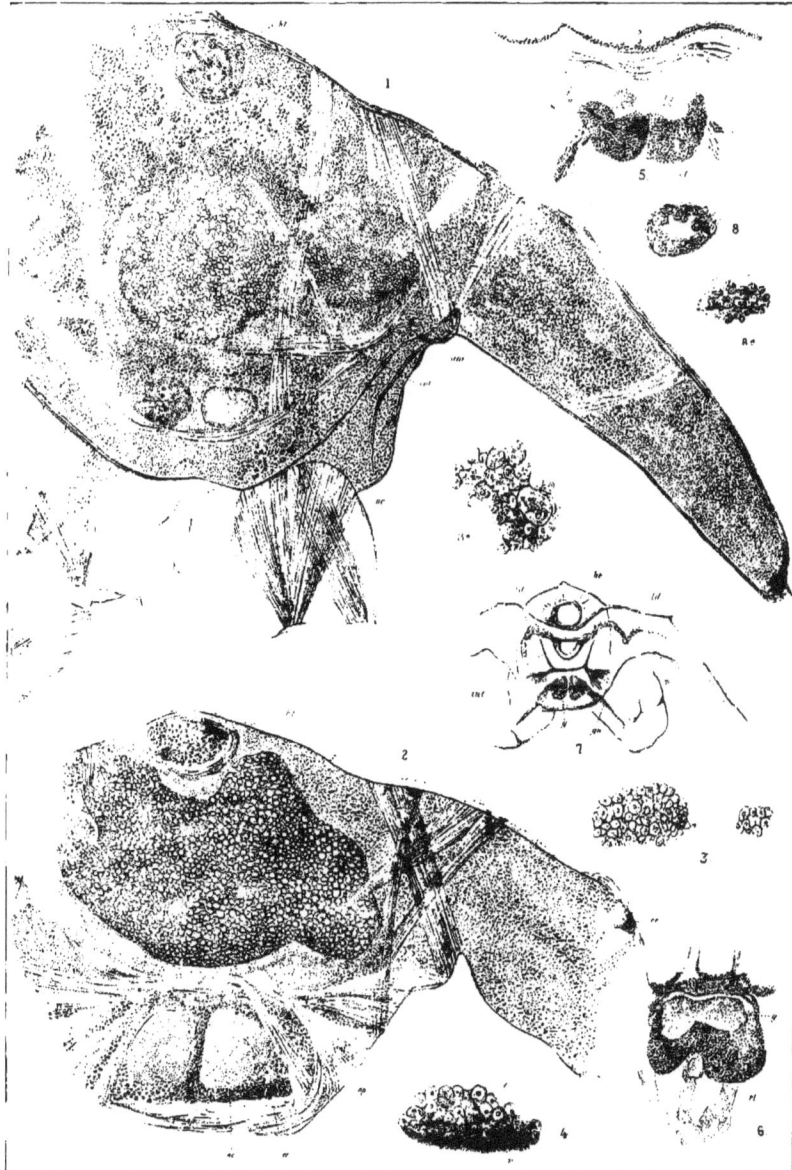

ANATOMY OF THE LARVAL KING CRAB

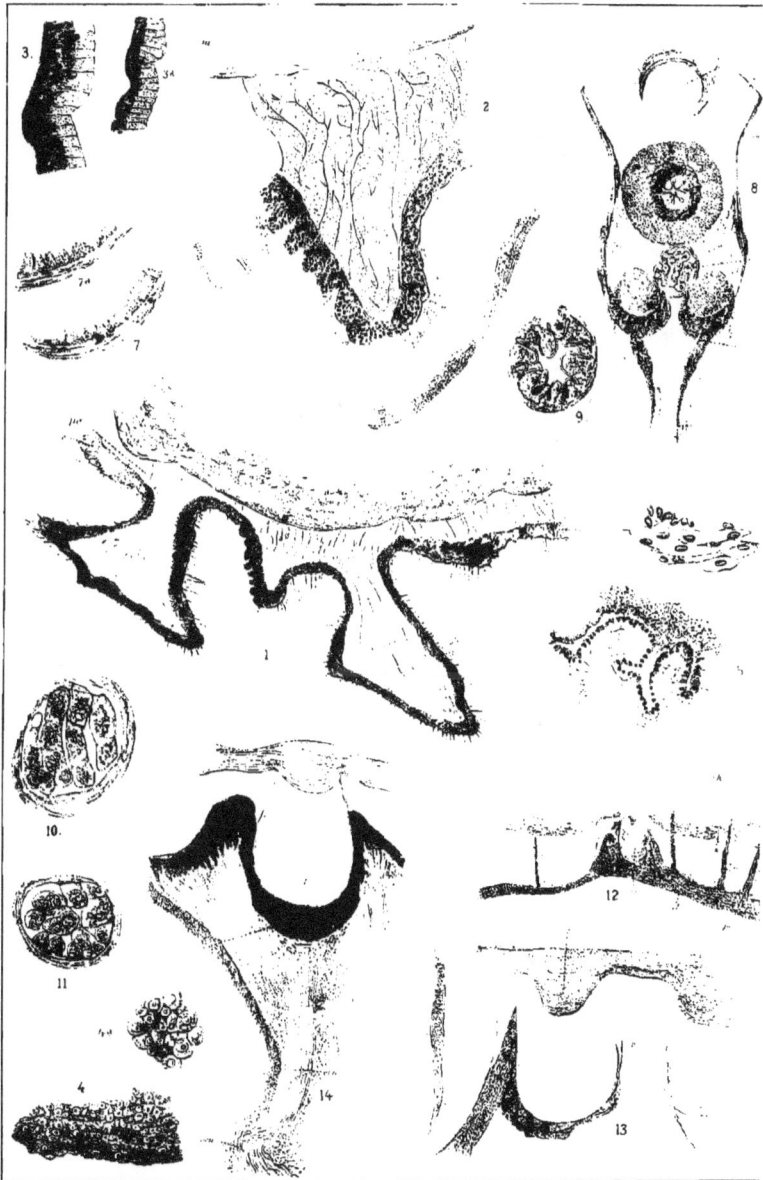

HISTOLOGY OF THE DIGESTIVE CANAL AND EYE OF THE KING CRAB.

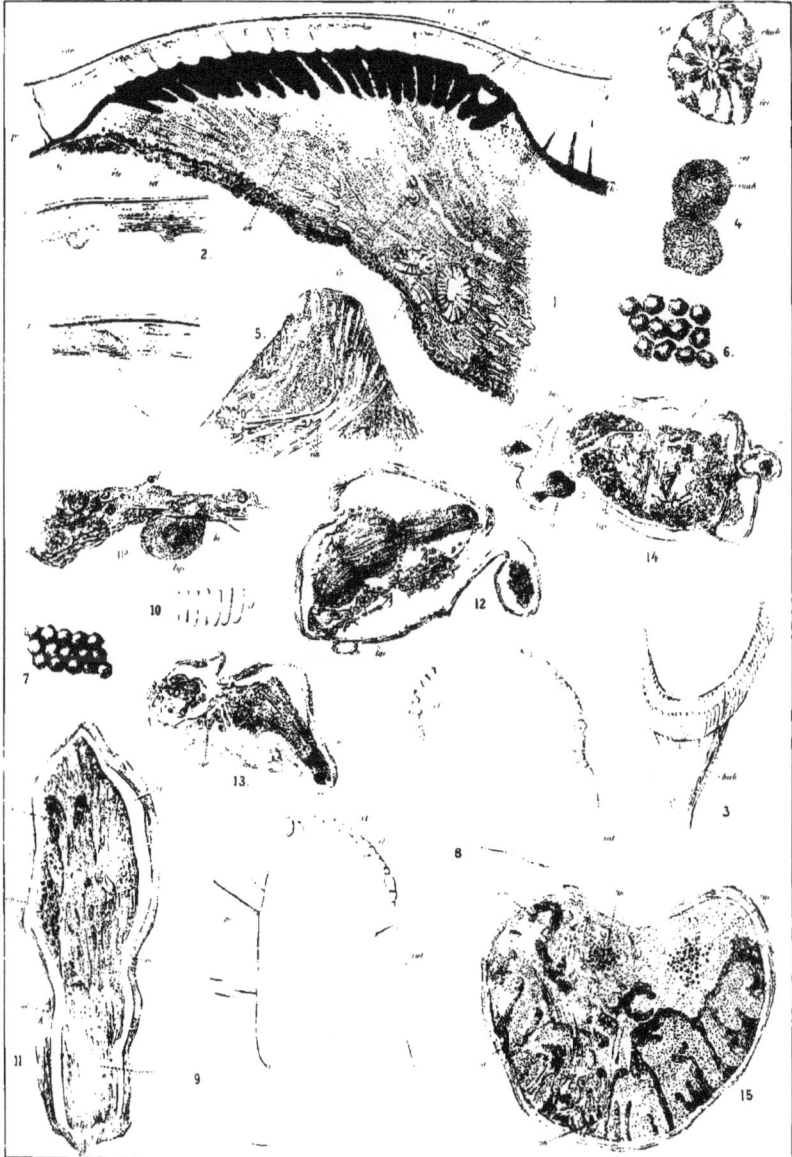

A. S. Packard del.　　　　　　　　　　A. Meisel lith.

STRUCTURE OF THE EYES AND NERVOUS CENTRES OF THE KING CRAB

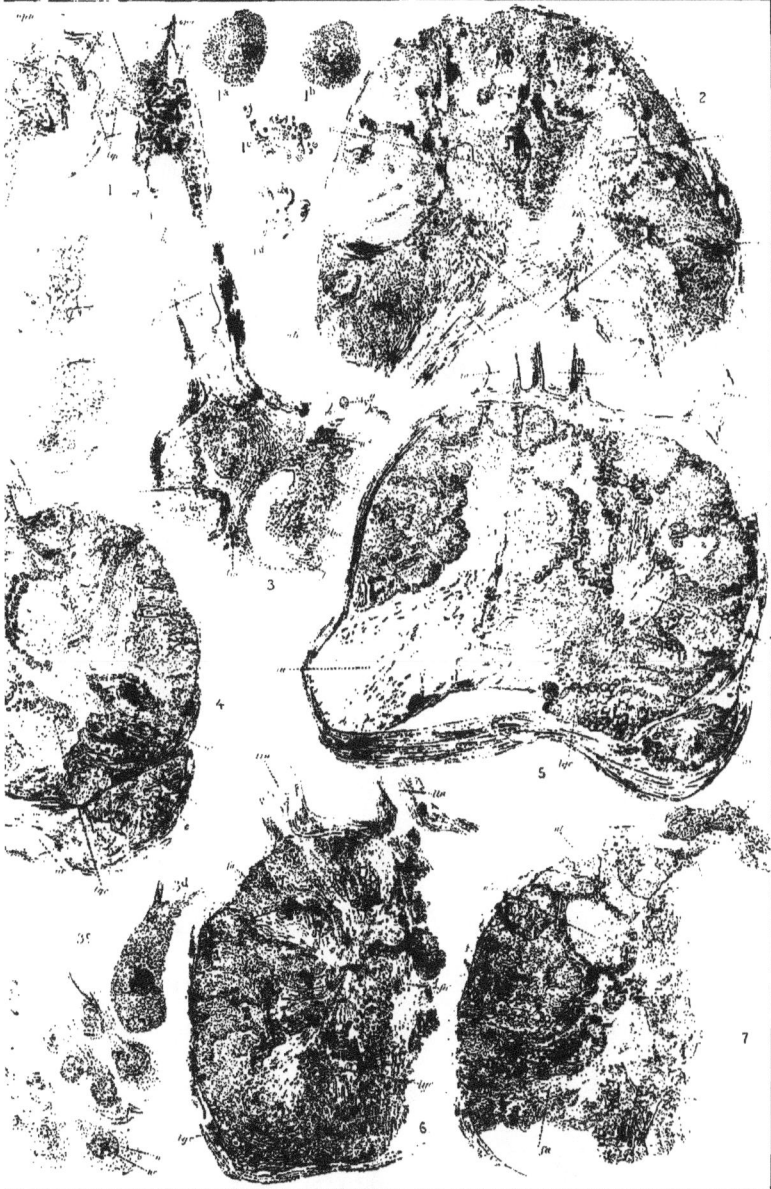

STRUCTURE OF THE BRAIN OF THE KING CRAB

www.ingramcontent.com/pod-product-compliance
Lightning Source LLC
Chambersburg PA
CBHW022014190326
41519CB00010B/1514